THE PROJECT

The past, present, and future of humanity

By Mark Macy

To Justine —
May your latest book
(I love the term 'sparkle brain')
bring Great Pleasures to
a troubled world.
Love & Light!
Mark Macy

Eloquent Books
New York, New York

Eloquent Books
An imprint of AEG Publishing Group
845 Third Avenue, 6th Floor – 6016
New York, NY 10022
www.eloquentbooks.com

ISBN 978-1-60693-749-5 1-60693-749-9

Printed in the United States of America

Book Design: Linda W. Rigsbee

To a lovely world too often trampled by confused, frightened, and angry feet. May caring hands and timeless wisdom soon lead you to your paradise destiny.

And to you, the special one—you know who you are—being groomed for your special purpose at this crucial time.

Table of Contents

Acknowledgements

FUNNY HOW WE'RE OFTEN unaware of those around us who wield the greatest positive influence on our lives and our world. So it is with invisible forces who've accompanied us individually and as a species for a long, long time, silently providing guidance and protection. And so it is with spiritual devotees whose focused will penetrates the veil to let love and wisdom stream to Earth. And so it is with the bright artistic and scientific minds around the world who grapple quietly with the world situation in a tireless quest for meaning and order. And so it is with restless activists locked onto a noble course, holding fast until humanity comes around. And so it is with the brave men and women of ITC research, and their invisible colleagues, contending with the dark cacophony to sustain meaningful dialog between our world and the finer realms of spirit through technical bridges. All of these good souls have my love and gratitude, as do my wife Regina and son Aaron who shine in my life like the brightest stars. And, of course, our good friends and extended family who provide safe harbor from the inner storm. God bless!

Introduction: Why We Are

SOMETHING MOMENTOUS HAPPENED IN our world at several points in our ancient past that forged the modern world and humanity. Understanding these mega-events gives us a clearer understanding of our world and ourselves and casts light on bright future scenarios that are part of our destiny.

In 1996 I was part of an international panel of scientists and researchers who were receiving, cataloguing, documenting, and analyzing what appeared to be a series of communications coming into our world from other worlds directly through our telephones, computers, TVs, radios, and other technical devices, as part of a futuristic field of research called instrumental transcommunication (ITC).* My colleagues in Luxembourg reported a lengthy contact from seven highly advanced beings through their telephone answering device. The message said, in part:

In the course of bygone decades, of thousands of earthly years, beings interested in human species meet to decide on the continuation of The Project.

* ITC (Instrumental Transcommunication) is the use of technology to get directly in contact with other levels of existence, what we often call the worlds of spirit.

For several years I assumed the message was referring to the project of high-tech, other-worldly communications that were starting to open up. That would have been astonishing enough. I eventually came to realize, though, that "The Project" revolves around humankind and goes back thousands, tens of thousands, maybe tens of millions of years, tracing to the very beginning of humanity on Earth.

If these ITC contacts are legitimate (and evidence has convinced me they are), then it would seem that many beings beyond the Earth are interested in the fate of our world, and they are coming to a decision. What is that decision? Why are they talking about us? Who exactly are they? And, most important for us perhaps: Who exactly are *we*?

Those are the subjects of this book.

PART ONE

Revisiting Our Ancient Heritage

WAS ATLANTIS A REAL civilization or just a myth? How about Eden, the legendary paradise world where humanity fell from grace? Real or myth? Three decades of research into humanity's true nature and heritage have convinced me that Atlantis and Eden were both real. While a few of the details in Part One may someday prove wrong, I'm confident that the overall story will bear out as future science discovers the truth about our prehistoric roots. Meanwhile, I hope the coming pages stimulate some deep-seated memories in you of ancient times that have been buried in human genes for thousands (if not millions...if not billions!) of years.

I've gathered a lot of information over the years from many sources and pieced it together in Part One into what I believe is a reliable picture of our world and what it means to be "human." My sources vary in reliability from impeccable to speculative, and as I assembled the puzzle, I tried to keep it all clearly delineated. It may surprise you how I determined which pieces of information are more reliable than others. Here in the high-tech world the general consensus among the media and general public (and, of course, the scientific community) is that scientists provide the most reliable information, but in my book they're only third on the list of four sources.

The two highest-level sources of information used in this book are not of this world. In 1995 I helped to assemble a panel for leading-edge spirit communication through technical equipment. We called ourselves INIT, the International Network for Instrumental Transcommunication. Our members—scientists and researchers from more than a dozen countries—received an unprecedented series of contacts from the worlds of spirit. Most of the contacts came from a group of people who had lived on Earth in various eras and had now come together on the other side to open a communication bridge to send messages to our world through ITC (Instrumental Transcommunication) systems composed of radios, TVs, telephones, computers, and other devices. I myself spoke on the phone and through the radio with spirit friends, as I've documented in my earlier books and on my websites*.

A few of the messages from the other side—what I consider to be the gems of ITC—came from Ethereal beings who have followed our world for thousands of years, offering humanity guidance and protection. Those contacts received by INIT members in Europe convinced me that under the right conditions, ITC systems allow information to be delivered from spiritual realms into our world almost pure, as opposed to channeled information, which is filtered by the minds of the human channels.

So let's look at my four categories of information sources†, starting at the top:

* For the most comprehensive ITC information available, visit www.worlditc.org.

†About the icons: The angels icon is a romanticized view of ethereal beings, who say, for example, that they don't really have wings . The double face is a picture of John Denver in spirit that I received via ITC systems. You can find a lot more about that and other such photos on my secondary website www.spiritfaces.com and in my book, *Spirit Faces; Truth About the Afterlife* (2006, Red Wheel/Weiser).

Ethereal truth. ITC messages from The Seven Ethereal beings, who identified themselves to us as "gatekeepers" between the spirit worlds and our world. They didn't reveal to us the full scope of their gatekeeper duties, but those duties certainly include control over the miraculous communication bridge that opened up for our group. Our spirit friends (deceased humans) told us that being in the presence of The Seven is like standing before a bank of living supercomputers that exchange oceans of information instantly. These entities provided vast energies to facilitate the miraculous contacts that our spirit friends made with us. Occasionally The Seven themselves spoke through the ITC channels, delivering messages to us through most of our communication technologies that boggled our minds, and I would trust the validity of those messages with my life (which, frankly, is a small risk, considering the timeless nature of the Ethereals and the fleeting nature of human lives). To say their information is as good as gold is an understatement. As a reliable source, then, messages from The Seven through ITC devices are number one in my book.

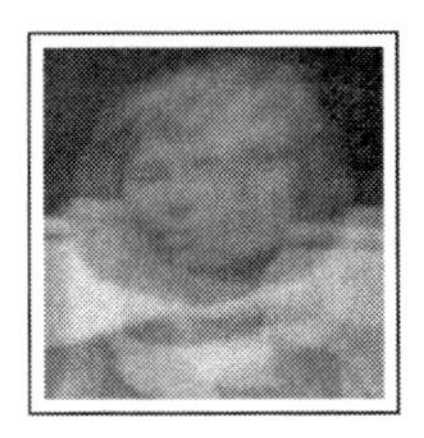

Spirit world insight. We quickly learned that people who die on Earth and resume lives in the Paradise world we often call Heaven or The Summerland, while retaining much of their mind, memory, and personality from lifetime, also have access to far greater understanding than we on Earth have at our disposal. Over the years it became evident to INIT members that the information our spirit friends shared with us—even information that our rational minds wanted to reject—usually proved to be right. So this book includes some of their information, which I find very reliable.

Hard evidence. Since the days of Isaac Newton, science has become brilliant at figuring things out, from the very large world around us to the very small world inside us. The scientific method of peer review, replication of results, and rigorous testing minimizes the risk of frivolous conclusions, but scientific understanding changes constantly as new discoveries are made. What is accepted in one decade can be proven false in the next. Also, scientific exploration today confines itself to the material universe, mostly ignoring what I regard as the much larger and more flourishing spiritual universes. So in terms of reliability, modern science is very good but has some limitations despite its prudence.

Soft evidence. Legends, channeled materials, religious texts, and speculation based on scientific evidence. At the source of many myths are kernels of truth that have been distorted in translation down through the ages. Tremendous wisdom can be found in time-proven religious texts (the Vedas, the Bible, the Koran, the Kabbalistic Zohar...) and esoteric writings such as *A Course in Miracles*, the *Urantia Book*, the Edgar Cayce prophecies, and the writings of Alice Bailey. But being channeled material, it was filtered to some degree by the beliefs and attitudes of the human channels as it came through into our world, and in some cases was modified by future generations. Hence, I call it "soft evidence." This would also include Plato's dialogs about ancient gods, and the works of serious writer-researchers such as Erich von Daniken, Immanuel Velikovsky, and Zecharia Sitchin, whose findings have not yet been accepted by the scientific community. These sources may not be at the top of my reliability list, but I hold them all in high regard. I believe their writings consist largely of truth but contain some misinterpretations as well, and so I'm selective in using that information.

You could say I tried to put Part One of this book together like a good suit: 1) Ageless ethereal truths from The Seven are assembled like well-cut, tasteful fabrics. 2) Spiritual insights from our invisible friends at Timestream pull things together like strong, durable threads. 3) Hard scientific evidence tailors the garment with precision to give it a good fit. 4) Soul-stirring legends and clever theories provide fringe and creative touches to soften the garment in a way that can set fire to eclectic minds.

So let's don the suit and set off on our adventure!

Note: Part One is written in an unusual style. Chapters 1-9 each start with a short hypothesis that tells a key part of the story of humanity and our world. The rest of each chapter supports the hypotheses with lots of evidence. So you have two choices for reading Part One. The first choice (which I recommend): First read all the hypotheses, the first paragraphs of chapters 1-9, to get the whole story...which may boggle your mind. Then go back and get unboggled by reading each chapter in its entirety, with all the supporting evidence, before moving on to Chapter 10. Or, alternatively, you can read all ten chapters straight through. Either way, enjoy!

Oh, one last thing: Reference materials are cited in this book in several ways, including endnotes (marked with small numbers), footnotes (marked with symbols), and *underlined italic passages* that can be googled on your home computer to find related information.

CHAPTER 1

Where Was Eden

My hypothesis. A planet called Eden (also called Marduk, Maldek, Phaeton, Faena, Tiamet, and other names) circled our sun in a regular orbit between Mars and Jupiter. It was a paradise world where all living things flourished in peace with each other, including the race of superhumans who served as stewards of that lush, perfect planet.

Ethereal truth. The paradise Eden, also called Marduk, was a world beyond the Earth where humans had no inclination toward "evil." [1]

Hard evidence. Today there are nine planets circling our sun in concentric orbits. Earth is the third planet out, Mars is the fourth, and Jupiter is the fifth. Between Mars and Jupiter is an unusually large gap, just the right size for another planet.

Soft evidence. When Babylon became the political center of the Euphrates Valley in the Middle East some 3,700 years ago, the god "Marduk" was head of the Babylonian gods, and he was believed to be associated with the planet Jupiter.

The Biblical book of Genesis[2] tells the story of God planting a garden in the east of Eden where people lived among the most beautiful foliage imaginable. When the people (represented by Adam and Eve) made bad choices, they were driven from paradise and replaced there by angels and a flaming sword to perpetuate "the tree of life."

(Over the centuries scholars deduced that if there was an Eden it was probably located somewhere in or around the Middle East. They overlooked the possibility of another planet—understandably, since space science is fairly new to civilizations of this Epoch. I'm convinced that the ancient writings were an effort to document actual happenings that were beyond human understanding at the time. Eden was a magnificent physical planet flourishing with life in a paradise ecosystem, in an orbit between Mars and Jupiter.)

CHAPTER 2

The Colonization of Earth

My hypothesis. The Edenites had mastered travel through space and among dimensions, and they colonized Terra (also called Earth), which was *not* a paradise world. Living things on Terra killed each other to survive in a ruthless environment. The Edenites regarded this world as a savage place, and they probably coined ancient words that reflected their deeply mixed feelings and that eventually evolved into modern words like *terrible*, *terrific*, and *terrifying*, referring to the savage Planet Terra. The Paradise People were trying to make this world a more hospitable place—importing paradise plants and animals of Eden to the colonies on Terra, and genetically engineering large, ferocious animals of this world to become small and docile. Their aim was to spread order in a world of chaos, but adapting to this wild world was difficult. They had to contend with the vicious creatures of Terra.

Ethereal truth. Long, long ago superhumans of Eden colonized the Earth, where they had to fight some of nature's most dangerous creatures that would not submit to human will.

Hard evidence. Dinosaurs roamed the Earth for some 160 million years, dying out 65 million years ago (and they were probably the dangerous creatures the Edenites had to contend with).

Soft evidence. In his 1968 book Chariots of the Gods, author Erich von Daniken suggests that intelligent extraterrestrials colonized Earth long ago and interceded in the emergence of humanity. There's ample evidence, says von Daniken, in the form of pyramids and other ancient monuments, monoliths, cave drawings, inscriptions, legends, and so on, that clearly suggest alien influences in the distant past.

More recently Zecharia Sitchin wrote eight books on the subject, including The Twelfth Planet and Genesis Revisited. He cites ancient texts (Assyrian, Babylonian, Canaanite, Sumerian, and others') that all refer to superhuman beings called Annunaki, Nephilim, and other names meaning such things as "Those cast down" and "Princely offspring" and "Those of Heaven and Earth." The superhuman beings landed around 432,000 years ago in the Persian Gulf, according to Sitchin, and they had come from a planet in our solar system. It was described in ancient texts as "a radiant world" (which I interpret to mean a paradise world—Eden, or Marduk).

CHAPTER 3

The Destruction of Eden

My hypothesis. Back on their home planet, the Edenites had technologies far beyond anything we have today. They could capture vast energies in crystals and send alternating currents through their planet. Eventually the energies reached critical frequencies that disturbed the planetary mass beyond the breaking point, causing complete and utter destruction. The planet Eden had circled our sun in a regular orbit between Mars and Jupiter until the advanced civilization blew itself up, and the planet became a hodge-podge of various-sized rocks and boulders that wrought havoc on the other planets and moons, and they still influence our solar system to this day. Much of the heavy debris pummeled the other planets and moons in the solar system, leaving them pock-marked with craters. Some of the boulders became asteroids that settled into orbit, forming a belt of debris around the sun between the orbits of Mars and Jupiter. Some were flung far into space and then were pulled by the sun's gravity into large, irregular orbits. They became comets. Smaller pieces, the size of pebbles and rocks, cluttered space within the solar system and became meteoroids, which when pulled into the Earth's atmosphere would burn up, becoming meteors that looked like shooting stars. Dust was spread throughout the solar system, some of it settling into rings around Saturn and other long-established planets. In short, most of the fragmentary heavenly bodies of varying sizes cluttering up our solar system today and over the past eons are nothing more than rubble left over from the self-destruction of Marduk, or Eden, and science has classified the bits of rubble as comets, asteroids, meteoroids, planetoids, rings, and so on, by size, location, and trajectory.

Ethereal truth. The planet Eden, also called Marduk, was inhabited by superhuman beings whose technology got out of hand and destroyed the planet in a massive explosion.

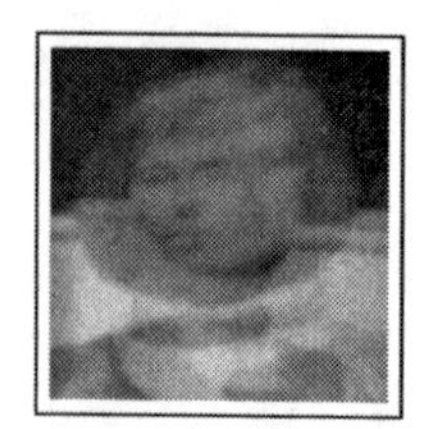

Spirit world insight. Most of our invisible friends at Timestream Spirit Group are humans who once lived on Earth, but since their death they inhabit a spirit world they call Marduk. Various groups on Earth have various other names for that spirit world. To Christians and Jews it's Heaven or Eden; to Muslims it's *Jannah*; to Hindus, *Pitraloka*; and to Spiritualists, the Summerland. Our spirit friends refer to it by the same name (Marduk) as the physical planet destroyed long ago.

(I'm certain that is not mere coincidence. Every material thing in the physical universe has counterparts in subtler realms, what I would call "spirit bodies." When a rock or tree or planet or human body is destroyed or dies in the physical universe, its spiritual template continues to exist in subtler realms. When we humans die, our spirit bodies live on, and I suspect that the spirit world Marduk is a template of the former physical planet. While all that's left of the physical planet today is an asteroid belt and other cosmic shrapnel, the spirit world Marduk is flourishing with life, providing home to many humans who once lived on Earth from many eras. It's still the radiant paradise that it had been in lifetime.)

Timestream sent us the following contact through a computer in Luxembourg in November, 1996. It was written by a woman in spirit who says she died on one of Earth's parallel worlds, a physical planet called "Varid." The contact warns of the destructive influence that certain kinds of scientific experiments can have on physical planets such as Marduk and Earth:

On planet Varid, our experimental team under the direction of Prof. Suat Dewar, pursued a theory that concerned the origin of matter and energy. It was based on experiments with a series of vacuum tubes which were contained within each other and would implode by themselves. It was our understanding that the total vacuum, or absolute void space, consisted of energy and "anti-energy". Sooner or later this would split, at least on planet Varid. (Unlike scientists on earth, we could produce a true artificial vacuum totally evacuated of air and matter to far below 1013 hectopascal) Energy and matter repelled each other, causing one such implosion of the tubes. We had come to the conclusion that our universe came into existence in two stages: 1. Splitting of the void into energy and anti-energy. 2. Transformation of energy into matter (suns and planets). We considered repulsion between energy and anti-energy to be the cause of the constantly increasing "drifting" apart of the universe. We now produced this anti-energy artificially, to spread all over our hermetically sealed laboratory in order to nullify atomic energy (also an artificial energy) which was also introduced in the lab. We were happy and confident that our discovery of "anti-atomic energy" would make impossible any atomic warfare, once we opened our lab doors and let it escape and spread on our world Varid, into the atmosphere and into space. Soon, however, it became clear to me that not everything was in order. Prof. Dewar, who was then in his 80s, noticed that his white hair started getting dark again. He did not complain anymore about his arthritis. It seemed to have disappeared. One of my female associates who was four months pregnant, started menstruating. Another younger girl complained about discontinuation of menstrual periods and shrinking breasts. All this pointed to a time reversal. I became panic-stricken when I realized what our experiments had triggered. The consequences would be disastrous if we could not reverse the flow of time back into its "normal" path. The anti-energy once flowing to the outside would reverse the aging process, but the living population would become so young that as premature babies they would be unable to withstand their environment and perish. Meanwhile no new babies would be born. Although a child could be conceived, it would not grow. It became obvious to me what we had to do. Under no circumstances should the locked chambers of our lab be opened. During our attempt to destroy our lab set-up a tremendous explosion occurred in which I died. It was 1987 October 30. I never found out what happened to Varid.... Dr. Swejen Salter.

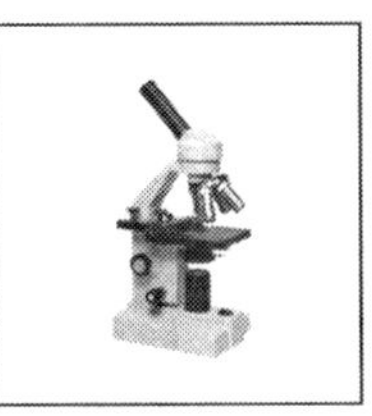

Hard evidence. In the gap between Mars and Jupiter, instead of a planet, there is an asteroid belt—a ring of rocks and boulders encircling the sun. Most scientists believe the asteroids are 4-billion-year-old debris left over from the formation of the universe, but through the years various insightful scientists—from Johann Bodes and Heinrich Olbers 200 years ago to astronomers Michael Ovenden and Tom Van Flandern today—have proposed a more reasonable idea: A massive, Saturn-size planet once filling that orbit was somehow destroyed, and much of the planetary debris settled into orbit, forming the asteroid belt. In 1987 scientists DP Cruikshank and RH Brown reported finding organic material (amino acids) and sedimentary clay on some of the asteroids, which could only have formed in water under the weight of gravity, as on an Earth-like planet.

Scientists today are convinced that the Earth and other planets endured megatastrophes in the distant past. The consensus is that our moon was formed when the Earth took a glancing blow from an even bigger object. According to recent findings, certain weird qualities of Venus—backward spin and earth-like composition but barren, bone-dry, and searing hot surface—suggest that it might have been the cosmic victim of an even more catastrophic, head-on collision. [3]

(Scientists have not adopted the notion of an obliterated Eden and its superhuman castaways on Earth playing a key role in the evolution of our world and the emergence of man. Not yet, but I'm sure they will, as it solves so many riddles about Heaven and Earth.)

Soft evidence. In May 2008 the Large Hadron Collider operated by CERN in Switzerland apparently experienced a phenomenon during the cool-down phase. Russian physicists who had helped in its construction reported an "*antiquark spree*" that hit the Earth's core and triggered a severe volcanic eruption in Chile, a 7.8 earthquake in China, and other geologic upheavals....

(This highly controversial report spread quickly around the Internet that month, and there's no telling at this point if there's actually any correlation between the CERN equipment and the earthquakes, but the grave implications suggest that we shouldn't be blithe about the possibilities.)

Some 100 years ago *Nikola Tesla* enjoyed experimenting with oscillators. He'd sit in his Manhattan lab and change the vibrations of the oscillators, and different items around the room would start to shake—maybe a chair, then a table, then a lamp. One night he strapped an oscillator to the thick steel pillar that ran up through the ceiling and down through the floor. As he adjusted the pitch, the pillar began to shake, then the entire room began to shake, and before long the vibrations streamed down through the pillar into the bedrock below Manhattan and set off an earthquake. Horrified, he smashed the oscillator with a hammer.

In 1912 Tesla was developing oscillator technologies that could send vibrations—sound waves—into the ground with earth-shaking results.[4] He said that using the planet as a carrier, his technology could deliver electricity anywhere in the world instantly if used appropriately. If used inappropriately, it could split the Earth apart like an apple. In February 1912, *World Today* ran an article by Allan L. Benson entitled "Nikola Tesla, Dreamer." An accompanying illustration showed the Earth blowing up, over the caption, "Tesla claims that in a few weeks he could set the earth's crust into such a state of vibration that it would rise and fall hundreds of feet and practically destroy civilization. A continuation of this process would, he says, eventually split the earth in two." (I believe Tesla was tapping into techniques similar to those that destroyed Eden.)

Zecharia Sitchin writes of a planet Tiamet that existed in the orbit between Mars and Jupiter, and another planet called *Nibiru* traveling in a huge oblong orbit that brought it close to the regular planets every 3,600 years, at which time there were geologic disturbances on the regular planets. According to Sitchin, a moon of the wayward planet Nibiru collided with Tiamet more than 4 million years ago, destroying Tiamet, forming the asteroid belt, and also forming the Earth, which had not existed before that collision. (I prefer the simpler scenario that I described earlier: The paradise planet Marduk (Eden), located between Mars and Jupiter, was destroyed by the advanced technologies of its superhuman inhabitants. As a result, the Earth and other established planets and moons in our solar system were bombarded by planetary shrapnel. If there is a Nibiru-like body circling the sun in an irregular orbit, it is probably a comet—one of the countless barren, wayward remnants of Eden's destruction.)

CHAPTER 4

When Did Eden Explode

My hypothesis. When science is able to analyze and date the asteroids in the belt between Mars and Jupiter, I believe they'll find that most of them were caused by a single explosion, and that will pinpoint the time of Eden's destruction. Meanwhile, the best we can do is to speculate, because we haven't received any reliable ITC contacts that state specifically when Eden exploded. They simply say, "Long, long ago..."

Craters might provide some clues. I suspect that many, if not most of the craters visible on the nearby barren planets and moons were created by debris from the explosion of planet Eden. It just makes sense that some of the impact craters on Earth, on other planets, and on our moon were formed shortly after the explosion, when debris was flying everywhere. If this is true, then our own world might have been colonized by superhuman beings—Edenites—much earlier than modern scientists would believe possible. The Ethereal beings told us explicitly that the Edenites had colonized our world *before* their world exploded. (We'll look at crater timeframes in a moment.)

Ethereal truth. Long, long ago when humans came to Earth from Eden (or Marduk), they lost mastery over nature.

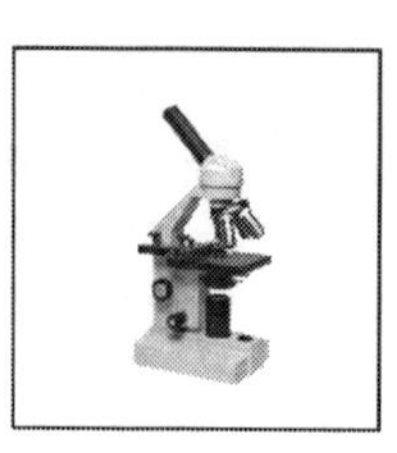

Hard evidence. The moon and nearby barren planets are covered by craters. Scientists until recently assumed that craters were formed by volcanic activity inside planets and moons, but thanks to the work of *Christian Koeberl*, *Virgil Sharpton*, *RAF Grieve*, *M Pilkington*, and others, today they have a better idea: As meteoroids fly through space, they sometimes collide with planets and moons, or burn and explode moments before collision, in either case resulting in craters. That's the prevailing theory today: Most craters are the result of impacts. Such impacts have played a big role in the evolution of our solar system, judging from the many pockmarks on the barren planets and moons.

Earth has also had many impact craters through the ages. Nearly 200 are known to science, but thousands more have almost certainly been erased or covered up by erosion and oceans and the spread of life on land. No doubt, many if not most of the lakes in the world were formed by rainfall gathering in old craters. Three of the best-known, scientifically recognized impact craters still prominent on Earth are:

- Wilkes-Land in Antarctica, a 300-mile-wide crater (the size of North Dakota or Cambodia), which was formed some 250,000,000 years ago by a 30-mile-wide meteoroid or asteroid,

- Chicxulub on Mexico's Yucatan Peninsula, a 110-mile-wide crater (the size of Vermont or Cyprus), which was formed some 65,000,000 years ago by a 6-mile-wide meteoroid or asteroid, and
- Barringer near Flagstaff, Arizona, a 0.75-mile-wide crater (the size of a small town or village), which was formed 50,000 years ago by a 164-foot-wide meteoroid.

So our world has been bombarded by meteoroids and asteroids, most of them hitting the Earth long before the first human-like apes are thought to have begun walking upright, which according to science occurred some 5,000,000 years ago.

The Apollo and Luna space missions brought moon rocks back to Earth in the 1970s. Argon-argon dating tests on the rocks suggest that the moon was bombarded by massive debris some 3.9 billion years ago, when scientists believe life was just forming on the Earth.

(If that lunar maelstrom was the result of the destruction of planet Eden, and if the dating methods are accurate, that would mean the Edenites were here studying our world at that time. And if life on Earth did indeed begin at the same time—3.9 billion years ago—then apparently the Edenites were here at that time not just to study life on Earth, but to establish it. That would mean that the Earth was to them a barren world that they converted to a garden. Either that, or else life on Earth was already thriving 3.9 billion years ago but was nearly obliterated by the explosion of Eden and had to be regenerated by the last living Edenites stranded here.

Let me suggest one other possibility—the most radical of all possibilities...but also, just maybe, the most accurate one: Time is an illusion of the physical world, and any attempts to measure time billions of years ago will give dubious results. As we study the very distant past with our preconceptions of time, we can alter the evidence with our focused intentions so that it conforms to our expectations. This is not necessarily something we do consciously with intent to deceive. More likely it occurs naturally—the result of a focused misconception—through a process that quantum physicists today are only beginning to explore, in which the observer affects the observed.)

Spirit world insight. Our main contact at spirit group Timestream has identified herself as Swejen Salter, a scientist from Varid, a parallel planet of the Earth. Science on Varid, she said in one contact, is much more advanced in ITC research than it is on Earth, and for that reason she was recruited into Timestream after she died in order to help with the development of ITC on Earth. In her role as Timestream director she shares with us mind-boggling information from the worlds of spirit that has made me reassess our notions of matter, energy, and time.[5] For example, she said that her world had once been flat at a time when everyone believed that it was flat, and today it is round, as everyone on Varid believes it is round. If her information is applicable to our world, it means that quantum physics is moving in the right direction, as consciousness plays an all-important role in shaping reality as we know it.

Soft evidence. Zecharia Sitchin says the superhumans first came to Earth some 436,000 years ago, but that the explosion of Eden (Tiamet) happened some 4 million years ago. (This doesn't jive with our ITC messages from the Ethereal beings, which state explicitly that the superhumans had colonized the Earth before their world exploded "long, long ago.")

In Peru, near the desert town of Ica, tens of thousands of stones have been found with ancient carvings showing people fighting dinosaurs, performing modern surgeries, and watching comets through telescopes. Others were etched with maps of the Earth—the way some people believe the Earth might have looked millions of years ago. The Ica stones are dark volcanic andesite, as hard as granite, ranging from the size of baseballs to the size of chairs. There are reports that around 1525 they were discovered by a Spanish Jesuit missionary, Father Simon, who traveled through Peru and asked about the rocks with strange animals engraved on them. Some samples were shipped back to Spain.

More recently the stones were collected by the thousands by Dr. Javier Cabrera, a retired physician, who purchased them from a local farmer, Basilio Uschuya. The stones can't be carbon-dated accurately by modern methods (they would have to contain organic matter for testing, which they don't). Still, the images themselves and the age-old finish permeating the grooves indicate that the carvings are very, very old, suggesting that advanced humans may have been alive on Earth more than 65 million years ago (when dinosaurs are believed to have died out). Scientists mostly ignore the Ica stones, assuming they're a hoax perpetrated by the farmer (even though some people estimate it would take more than sixty years for someone slaving everyday to carve 25,000 of them with a modern dentist drill). Uschuya generally has denied forging the stones, although on some occasions he admitted forging them, perhaps because he was arrested by Peruvian police for selling historic artifacts, a serious offense in Peru. By claiming forgery he could continue to sell the stones. So the farmer's mixed claims provide ample reason for scientists to dismiss them as a hoax, but I suspect the carved Ica stones are genuine relics from our very distant past—at least 60 million years ago.

Researcher Ed Conrad has discovered other inconvenient evidence that humans were alive at least 280,000,000 years ago, maybe earlier. He discovered what seem to be petrified human bones and organs between veins of anthracite coal in Pennsylvania. Coal was formed in the Carboniferous period some 280,000,000 years ago, according to scientists, which means that the original owners of those bones and organs must have lived and died at that time. Again, scientists for the most part ignore Conrad's claims, and some researchers argue that what Conrad really found is just a set of iron nodules that bear some resemblance to bones and organs. So the jury is still out on Ed Conrad, but if his radical claim proves true—that human bones were fossilized in the Carboniferous period—then it would mean that humans were alive on Earth a quarter billion years ago!

In June 1968 William J. Meister, a teacher and amateur fossil collector, found what seems to be a 600-million-year-old fossil of a sandaled human foot stepping on an ancient trilobyte in what is now Utah. Scientists reject the rock as a sandal-like impression created by natural causes, but the sandal resemblance is uncanny. If Mr Meister's claim is true, then humans were walking the Earth 600 million years ago!!

Researchers Michael Cremo and R Thompson wrote about hundreds of 2.8 billion-year-old grooved spheres, about an inch (2-3 cm) in diameter, dug up in South Africa in the 1980s. Although science has since offered possible natural explanations of the spheres that were found in age-old Precambrian strata, the small items certainly have the appearance of being manufactured. Time will reveal the actual origin of the spheres. Meanwhile, they do provide compelling soft evidence of human habitation of Earth 2.8 billion years ago.

So Eden might have exploded (leaving advanced humans stranded on Earth) 65 million, 280 million, or 600 million years ago, or 2.8 billion years ago, according to the work of Cabrera, Conrad, Meister, and Cremo. Various other research suggests other timetables, so the debate will rage on in the coming decades. Meanwhile, it's probably wise not to draw hasty conclusions about specific timetables until after the asteroids between Mars and Jupiter have been dated.

How many years ago did Eden explode, stranding colonists on Earth	
50,000?	The Barringer crater in Arizona was formed then, probably by debris from the explosion of Eden, but long, long after the explosion.
436,000?	Colonists arrived on Earth, according to Zecharia Sitchin's speculations. (More likely they arrived much earlier.)
5,000,000?	According to science, upright humans first began to walk the Earth at this time. (More likely that happened much earlier.)
65,000,000?	Dinosaurs died out around that time, when the Chicxulub crater in Mexico was formed, again, probably by debris from the explosion of Eden, but maybe long after the explosion. Ica stones etched with detailed pictures of people and dinosaurs, if legitimate, could date back to that time.
250,000,000?	The Wilkes-Land crater in Antarctica was formed then, probably by debris from the explosion (but the debris might have been floating in space for millions of years before colliding with the Earth).
280,000,000?	Human bones may have been fossilized in coal dating from the Carboniferous Period.
600,000,000?	There's a fossil of that age that resembles a sandaled footprint with imbedded trilobites that are known to have existed at that time.
2,800,000,000?	Grooved, metallic spheres of that age were dug up in South Africa in the 1980s.
3,900,000,000?	Rocks of that age found on the moon in the 1970s could be debris from the explosion of Eden or from a different big cosmic event.
Or . . . ?	Time is an illusion of the physical realm and becomes nebulous in relation to events that far in the past.

CHAPTER 5

The Marooned Edenites

My hypothesis. Superhuman colonists from Eden were on Earth at the time Eden exploded— which I suspect was 3.9 billion years ago, but might have been later. Scientists today feel pretty certain that momentous things were underway between 3.9 and 4.5 billion years ago, including the formation of the Earth and the emergence of life. I believe that one of the most momentous events at that time was the destruction of Planet Marduk, or Eden. Again, I feel fairly confident in that time-frame, though of course I'm not certain of it. In any case, the colonists became castaways here when their home planet exploded.

Eden had been a paradise world inhabited by an advanced race of humans whom we today might regard as superheroes. I suspect that fanciful stories such as Superman (who became stranded on Earth with superhuman powers after his native planet Krypton exploded) are the products of ancient memories bubbling up as inspirations in receptive minds of writers and story-tellers. And that's why such stories and legends are popular among people everywhere; they stir up our own ancient, long-buried memories as we read books or watch movies about them.

I suspect that by our standard of time the Edenites were virtually immortal, because they could move back and forth between the physical realm and timeless subtler dimensions—what we sometimes call the worlds of spirit. A multidimensional existence freed them from the rigors of aging, allowing them to move from subtler dimensions back into this dense physical universe with vital bodies in peak health at the prime of life. To humans like us who are stuck in this physical domain in a relatively short cycle of birth, old age, and death, these multidimensional humans would appear to have life spans of thousands of years. They would be superheroes. Titans. Gods. Annunaki. Nephilim. Kachinas....

If Earth was a very unpleasant place for the stranded Edenites, as I believe it was, then I suspect they would have chosen to spend less and less time in the physical realm. They would have begun spending most of their time in subtler dimensions as spiritual beings and less time as physical beings. Perhaps the Edenites are still with us to this day, living in subtler realms, monitoring our world; inspiring us in our dreams, daydreams, and meditations; and "down-modulating" themselves from time to time to enter the physical domain in order to partake in the affairs of our world. Whether they would appear to us as angels or advanced extraterrestrials or highly gifted men and women or prophets or saviors, or some combination of these...your guess is as good as mine at this point!

My research and experiences over the past twenty years suggest that all humans are multidimensional beings, our existence as purely physical beings with fleeting lives is just an illusion in which most of us are trapped, and we can learn to break free of the illusion—to perform feats that seem miraculous and to get in touch with our true, timeless self—as we foster our multidimensional nature through prayer, meditation, and other spiritual practices. There are people in the Far East living such lives, and we could learn to do the same.

Ethereal truth. The superhuman colonists from Eden became marooned on Earth when their home planet was destroyed, and they were the last living Edenites.

On March 4, 1996, our INIT members in Luxembourg received a computer contact from The Seven Ethereal beings. They referred to "Titans," powerful superhumans of Greek mythology, though I suspect those Titans they were referring to were not really mythological. They told us in part:

Though you (INIT) are still a small group, by your publications, meetings and lectures you can change the world and keep the door to other dimensions open, even in dark times. We, the Seven, have a rational, instinctive intellect, very different from yours. We know each of you in INIT and know your weaknesses. We understand you and love you in spite of your weaknesses. In this life you will always remain children with the dreams of Titans. What is so bad about that? We too do not claim to know everything, although our wisdom and knowledge seem to be unlimited next to yours…

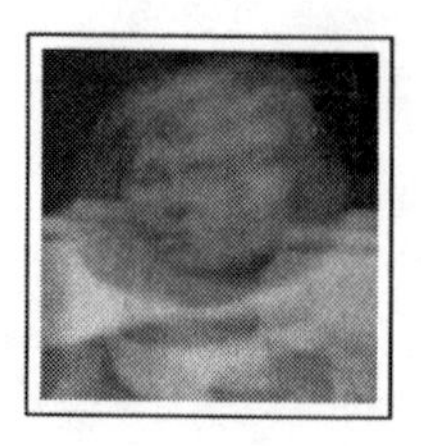

Spirit world insight. Our spirit friends at Timestream Spirit Group told us that when traveling to other dimensions they have techniques called "light modulation," which moves them to subtler realms, and "down modulation," which carries them to denser worlds. In this way they can visit the Earth, for example, taking on bodies dense enough to affect our equipment, and then return to their paradise world where their bodies become light again...or they can travel to dark, dismal realms to find lost souls—people who have died and gotten stuck there—and accompany them to the paradise world to be rehabilitated. (I suspect the Edenites had mastered similar techniques to help them travel to subtler realms, and then return to the very dense physical realm.) Here on Earth, psychics often talk about lost and troubled souls "going to the light." The light is paradise. It's the paradise existence vibrating at a fine rate that's blinding to those of us on Earth lucky enough to get a glimpse of it. So when people from the dark realms are brought to the light, it takes some adjustment.

Soft evidence. There are said to be mystics living in the Himalayas today leading multidimensional lives, retaining their youthful physical form for centuries. Mahavatar Babaji is one such ageless ascended master who inhabits the Northern Himalayan caves near Badrinath Temple in northern India. Mahavatar Babaji appeared in front of Lahiri Mahashaya between 1861 and 1935. Lahiri Mahashaya, after being initiated into Kriya Yoga by Mahavatar Babaji, then became the guru or teacher of Sri Yukteswar Giri, who also described his encounters with the ascended master Mahavatar Babaji in his book, *The Holy Science*. Sri Yukteswar was, in turn, the guru of Paramahansa Yogananda, who also discussed Mahavatar Babaji in his own book (one of my all-time favorites), *Autobiography of a Yogi*. Yogananda writes, "The Mahavatar Babaji is in constant communion with Christ; together they send out vibrations of redemption, and have planned the spiritual technique of salvation for this age. The work of these two fully-illumined masters—one with the (physical) body, and one without it—is to inspire the nations to forsake suicidal wars, race hatreds, religious sectarianism, and the boomerang-evils of materialism."

The ancient Greek philosopher Plato wrote that very early on, the Edenites (whom he called "gods") divided the inhabitable Earth among themselves as stewards of various districts, providing wise, gentle guidance to the native humans. Plato had learned about the ancient gods indirectly from the Greek statesman *Solon*, who had learned about them from the Egyptians, who in turn had inherited their knowledge from earlier civilizations. Most of the details of our ancient history were lost during long dark ages when people were preoccupied by survival, Plato said. During those difficult periods, few facts were preserved accurately, so that by the time of Plato the ancient stories sounded more like myth than history. The only details that were preserved accurately, said Plato, were the names of the gods. Athena and her brother Hephaestus loved philosophy and art and were in charge of the area that is now Greece. Poseidon was in charge of Atlantis. Other Edenites were in charge of other regions. The Earth was a violent world, and so the human cultures that evolved under the stewardship of the Edenites included warrior classes as well as farmers and craftsmen.

CHAPTER 6

Cross-Breeding

My hypothesis. The stranded Edenites wanted to seed our world with a new species of humans, and I suspect they did a lot of genetic engineering to come up with human beings bright enough to reach a high level of decency and understanding, and rugged enough to survive the rigors of this wild planet Earth, where animals fought ruthlessly and killed each other for food and mating rights.

Did the Edenites cross-breed with primitive, animalistic humans native to Earth after they arrived here and became stranded, or did they genetically engineer terrestrial humans from the start? At this point, I'm not sure about that, but I *am* confident that the cross-breeding or genetic engineering did indeed take place. I'm beginning to believe that the Edenites were busy 3.9 billion years ago cultivating life on Earth (and probably on Mars and Venus too) when their home planet was destroyed, and life on Earth was the only physical life in our solar system to survive the cosmic cataclysm.

In any case, we today are products of that ancient cross-breeding. And that's why we all have both a noble side that yearns for a life of order, love, decency, and good will, and a savage side that pulls us into wild dramas stirred up by our fears, insecurities, cravings, and animosities. (More about that in Part Two.)

There were many other races of human cross-breeds engineered before we were—not only such well-known ape-like species as Australopithecus and Homo Habilis, but also lesser-known races of god-like humans, giants and dwarfs, who appear in ancient stories from around the world. Evidence abounds in anthropology textbooks of primitive human species thriving in ancient riverine and coastal communities in various locations, but little more than a footnote is devoted to giants and little people and gods. These are among the troublesome truths that don't fit the scientific worldview, so they're hidden away in museum closets and dusty archives of scientific institutions. In any case, it was probably the powerful races of giant human cross-breeds who built massive stone monuments at Stonehenge, Easter Island, and elsewhere, and it was probably the marooned Edenites and/or their direct descendants on Earth who built the pyramids in Egypt with technologies carried over from the mother planet.

"Homo sapiens sapiens" (the modern human being) appeared on Earth some 30,000 years ago or 90,000 years ago or 130,000 years ago (depending on whose theories and evidence you prefer). I suspect the first highly successful generations of our ancestral hybrids displayed some of the superior intelligence from the Edenite side of their family tree, but also much of the fear, insecurity, cruelty, and urge to dominate and destroy that came from their wild terrestrial side. Scientists believe that these wild compulsions evolve naturally among animals that have to adapt to a fight-or-flight, kill-or-be-killed ecosystem, and that is probably true. Our savage qualities make survival possible in the wild, and coexistence difficult in society; and they characterized the nature of the earthly humans long ago. Our mixed nature is probably the result of both evolution and ancient genetic engineering. As a result, the noble-savage cross-breeds became at best clever, adaptable, and unpredictable, and at worst the scourge of the Earth for a long, long time as they used their cunning and intellect to beat wildlife into submission. Even to this day, our noble side urges us to be wise and gentle stewards of the Earth while our savage side compels us to venture out into the wild with shotguns (as hunters) or with rocket launchers and machine guns (as soldiers) to kill food or to conquer the ruthless, incorrigible forces around us. These savage tendencies continue to drive us humans in modern times, when we no longer have to hunt for food and when most of those ruthless forces are long gone from our world and exist only in our minds. Old patterns are hard to break.

We're all in this together. When we buy meat from the market bins or a burger at a fast-food restaurant, we perpetuate another savage reality: Millions of calves are castrated, grazed for a year, then fattened up in feedlots where they wallow in their own waste for several months and eat from troughs until the day when they walk with their friends, screaming in terror, into slaughter-houses, fully aware of the grisly reality awaiting them at the front of the line. Not a pretty picture.

Hunters, soldiers, and butchers need not feel any more or less guilty than the rest of us, though. The fact of the matter is: Living things on Earth kill each other to survive. It's the natural cycle here, and we're part of the cycle.

It's often frustrating to be human. Our savage side accepts brutality as a fact of life, while our noble side shrinks in horror from the agony of innocent creatures around us, much of it imposed upon them by us. We sometimes find curious ways to alleviate the guilt—pretending that the lives of nonhuman animals and plants are unimportant, or that they can't really feel pain, or convincing ourselves that a human fetus is all-important while millions of children, women, and men who die each year of starvation and war can be ignored. We all have ways to cope with the frustrations of living in a brutal world, and we'd probably get along better if we admitted to ourselves that they are coping mechanisms rather than defending them vehemently, often violently as righteous principles.

Ethereal truth. (At some point) the super-human castaways from Eden developed a civilization called Atlantis, and they crossbred with the primitive natives on Earth, resulting in legends such as the Greek Titans and man's fall from grace. The ancient cross-breeding also led to modern humans.

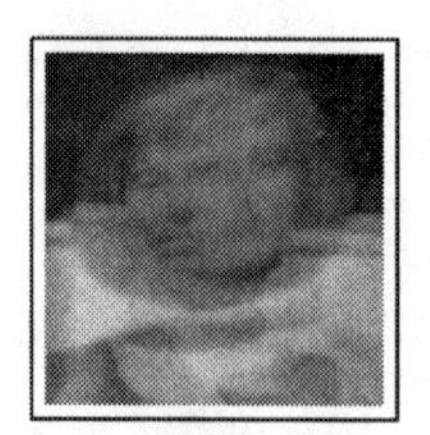

Spirit world insight. The success of our international ITC group INIT from 1995 through 1998 hinged on what our spirit friends called a "contact field," which was a pool of all the thoughts and attitudes of everyone involved in our ITC project on both sides of the veil. The field was invisible and abstract to us on Earth, but our spirit friends said they could see and feel it. When the attitudes of everyone involved in the project were mostly in harmony, the field was clear, and our spirit friends could feel a closeness to us all; they could see into our world and work with our equipment. But when there was dissonance in the form of chronic envy or fear or doubt or resentment, the field became cloudy, and it became very difficult for them to work with us. They began to lose touch with us because our vibrations no longer matched theirs. It was like radios going out of tune with each other.

When dissonance began to prevail in our group around the year 1998, the contacts from our good spirit group began to wane, and negative spirits began to break in through our equipment, giving us unsettling, sometimes frightening messages. My experiences with INIT made it clear that there were light and loving spirit worlds whose friendly inhabitants supported our efforts, and there were dark and dismal realms where spirits entered our world only to stir up our troubled thinking.

Those heavenly and hellish spirit realms trace back to the ancient cross-breeding of humanity that produced the noble-savage humans. For hundreds of thousands of years, as our noble side emanated thoughts and attitudes of love and good will, those fine vibrations streamed outward (or inward, depending on how you look at it*), creating many of the realities in the paradise world, and the light vibrations in paradise fed back into our world. It's been an on-going cycle of love and light feeding on itself and growing.

* We'll discuss the actual location of the spirit worlds in Part Two.

At the same time, our savage side has been sending out dark vibrations of fear and animosity that stream into dismal spirit worlds to feed the darkness. The troubled thoughtforms reenter our world to stir up hatred and fear throughout society.

Humanity has passed through many cycles in which one side of human nature prevails for several decades or centuries, and then the other side prevails. As that happens, the light and dark spirit realms ebb and flow, accordingly, in terms of their influences on our world. We flourish in eras of enlightenment and suffer through dark ages. Humanity is now in a period of swelling darkness, possibly leading to the end of an Epoch, or an "End Time."

Hard evidence. Life in the rough terrestrial ecosystem isn't easy, so the body has been engineered, whether through evolution or ancient cross-breeding (probably both), with a built-in pleasure drug. The hormone dopamine is produced by the body to give us a surge of pleasure through a "reward pathway" in the brain and nervous system when we do things that perpetuate survival—such things as eating, sleeping, and having sex. Recent studies by Vanderbilt University graduate student *Maria Couppis* indicate that our body rewards us for aggression too. She found that caged male mice activate triggers to have fighting partners put into their cages with the same eagerness that they'd exhibit to get cheese or peanut butter.[6] Since mice and men are behaviorally and genetically similar, such studies make it clear that our lives on Earth come with a package of animal emotions that most of us spend much of our lifetimes having to contend with. They help explain why many of us are attracted to violent sports, movies, and pornography, and why TV news programs focus on crimes, disasters, and violent demonstrations; they attract viewers by stimulating people's reward pathways.

(As that scenario unfolds on Earth, the darker spirit realms swell, and they have ever-greater influence on our world. And, again, I believe it goes back to a time tens of thousands of years ago when the god-like humans from Eden conducted genetic engineering experiments to come up with rugged, adaptable humans who could flourish in the raw terrestrial environment. It was probably those ancient experiments that programmed our brains to "survival mode," complete with savage side as well as noble side.)

Roman mythology centers around the story of Romulus and Remus, hybrids of the god Mars and a vestal virgin human girl named Rhea Silvia. The twin brothers are said to have founded Rome in 753 B.C. While much of the story might be embellishment (for example, that the twin babies were abandoned and suckled by a she-wolf), recent evidence uncovered by Italian anthropologist Andrea Carandini suggests that the crux of the story, at least, might be based on historic fact. In 2007 he discovered the remains of a royal palace dating back to the time of the city's legendary beginning, at the precise spot where the Temple of Romulus stands today.[7] There may be some truth to the stories of god-human hybrids Romulus and Remus, and the founding of Rome.

Based on fossil records, Neanderthals were rugged humans—hairy and thick-boned with less reasoning power than we have today—roaming the Earth nearly a half-million years ago. They were the dominant species until humans similar to us appeared in Africa 130,000 years ago. By some accounts, something strange happened some 30,000 years ago. Modern humans called "Cro Magnons" suddenly appeared "out of nowhere" in Europe and began to coexist with the Neanderthals. There's no solid evidence that the newcomers, with their more delicate bone structure, upright posture, disappearing body hair, and heightened sensibilities, had evolved from any of the primitive human or nonhuman predecessors of this world. The Cro Magnons were something new to the Earth when they simply appeared, (and I suspect that they were—and we are today—the result of cross-breeding or genetic engineering by the Edenite superhuman castaways on Earth and their gifted descendants in Atlantis.)

Remains of 3-foot-tall (1 meter) people were discovered in 2003 in a cave on the Indonesian island of Flores. Called *Homo floresiensis* and nicknamed "*Flores Hobbits*" by researchers, the little people had intelligent human-like brains proportional to their size. They appear to have died out some 12,000 years ago.[8]

Many secrets to our past have been hidden away in caves and ancient lake and river beds, in lava flows and burial mounds from distant cultures. Old burial mounds were prevalent in many countries 200 years ago, before cities, highways, and megafarms flattened everything out. Many 19th Century amateur excavators, farmers, miners, and construction workers dug up skeletons of giant humans. These are not just a few isolated reports of *giants in history*, but dozens of them from all around the world. Ample stories and pictures can be found on the Internet.[9] Some reports are obvious hoaxes, but others are quite credible. So giants played an important role in the human drama over thousands of years, and stories like the Biblical "David and Goliath" are probably rooted in fact.

Soft evidence. 2,000 years ago, Jewish historian *Flavius Josephus* wrote, "There was a species of giants, who had bodies so large, and countenances so entirely different from other men, that they were stunning to see, and terrible to hear. The bones of these men, still displayed to this very day, are unlike any credible relations of other men."

In the 18th Century BC the Babylonians believed that gods living on Earth long ago became divided, leading to a war between two factions. The Annunaki gods met to select a leader, and Marduk, a young god, rose to the occasion. Marduk defeated the opponents (the goddess Tiamet and her army), "he wrested from the enemy the *Tablet of Destinies*, wrongfully his," and he became leader of the Annunaki. Under the reign of Marduk, humans were created to perform the difficult and mundane tasks of surviving on Earth, so that the gods could attend to more important matters.

According to Zecharia Sitchin, the colonists from space spliced their own genes with the genes of Homo Erectus to create Homo Sapiens (modern man).

Plato suggested that Poseidon was one of the original Edenites. He wrote: "Poseidon, receiving for his lot the island of Atlantis, begat children by a mortal woman, and settled them in a part of the island...." According to Plato, the young woman, Cleito, bore five pairs of twin boys for Poseidon, the first-born being Atlas. Poseidon's large island was eventually subdivided among his ten sons, and then among his grandsons and great-grandsons.

In Genesis 6 of the Bible, as human population grew, the Sons of God (divine beings, or superhuman Edenites) began to pick the most attractive of the human women and marry them. Their progeny became the heroic men and women of legends from various cultures.[10]

(Many of our myths about ancient human-gods seem to be based on fact; there was a lot of cross-breeding going on between the primitive humans of Earth and the superhuman castaways from Eden. However, it seems unlikely that a god-like human would find an ape-like human of the opposite sex attractive enough to have sex and raise children. It's more likely that god-like hybrids such as Atlas, Romulus, and Remus were the result of genetic engineering, and romantic myths of intermarriage with the gods were "humanized" accounts of actual events. As centuries passed and myths spread, the primitive early humans could probably understand and identify with stories of sex, marriage, and child-rearing, but not genetic engineering, so the myths of divine intermarriages proliferated. Even so, some early writings seem to make an attempt at describing genetic engineering, as found in the Bible, below.)

In Hebrews 11, the prophet Enoch was visited by angels (very tall men with faces that shone like the sun, eyes that glowed like lamps, wings of gold, and hands white as snow), and they took him on a tour of the spirit worlds. First stop was an Ethereal world of angels and light beings. Then he was taken to a dense, dismal world where lost souls and troubled spirits implored

Enoch to pray for them. Then Enoch was shown an astral paradise called Eden that was sustained by a magnificent tree of life. Onward, then, to another ethereal realm inhabited by angels and light beings. Then to another dense spiritual world inhabited by godless, remorseful beings called "Grigori" (perhaps what we today would call "reptilian extraterrestrials") who had lost their connection to God, and finally to a highly refined ethereal world of archangels who were so majestic that they frightened Enoch. The wisest of the archangels spent a month teaching Enoch all about the vast worlds of spirit and the human role in the big picture, including humanity's fall from paradise. Enoch recorded the information as accurately as he could in a series of books and was told to give the books to no one, but to share the information with anyone who wanted it—to spread it as far and wide as possible. Because of Enoch's ethereal connection, his family was blessed for several generations. (I believe the blessings came in the form of genetic cross-breeding by highly advanced beings to produce Enoch's progeny.) His son Methusaleh (whose name meant "after his death a flood will come") lived to be several centuries old. Enoch's grandson Lamech had a wife who bore a child with blond hair, white skin, and illuminated eyes that filled dark rooms with light. Bear in mind that the early Hebrews were a swarthy lot, so a fair-haired, light-skinned baby with glowing eyes would have been startling and out of place. That great-grandson was named Noah, and when he grew up he built the proverbial Ark to survive the Great Flood when his grandfather Methusaleh died. Was Noah an albino? He was probably much more than that if his eyes filled dark rooms with light!

(The Great Flood was probably perpetrated by the powers-that-be as the First Epoch of humanity came to a close amid moral decay and ruthless barbarism. It was an attempt to wash away the traces of a great civilization gone bad and to start the Second Epoch with a clean slate… but I'm getting ahead of myself.)

CHAPTER 7

Atlantis

My hypothesis. The so-called "Cro Magnon man" appeared in Europe some 30,000 years ago—the noble-savage product of genetic engineering by the marooned Edenites and their gifted descendants here on Earth. Atlantis was a great civilization, probably established several millennia after the introduction of the new humans. It was built with some of the advanced technologies left over from Eden, the mother planet, and it marked the beginning of the First Epoch of modern humanity.

The Atlanteans were a war-like civilization, looting less advanced civilizations in the Mediterranean area. Their empire covered most of Europe, the Middle East, and Northern Africa. The capital—at least in the final phase of Atlantis—was near the present-day island of Helgoland off the coasts of Denmark and Germany.

Atlantis flourished for thousands of years and disappeared much more recently than most people today might guess, bringing the First Epoch to a close. As recently as 2400 BC the Atlanteans were transmitting massive energies through the Earth from immense crystals, causing powerful geologic upheavals throughout Europe and the Middle East. Those earth-shaking experiments went on for hundreds of years, causing widespread destruction. The last and worst explosion around 1220 BC caused what remained of the Atlantean civilization to sink into the ocean. On the mainland of Northern Europe, legends of Norse gods emerged from the rubble, and the war-like Atlantean legacy was inherited by the Viking culture.

On a related note, I would not be surprised to learn that certain Tesla technologies have been gently inhibited during the past century by forces beyond our world—what are sometimes called "guardian angels"—to prevent humans today from repeating the catastrophic mistakes of Planet Eden and the civilization of Atlantis.

Ethereal truth. Legends of man's fall from grace are based on an actual incident—the downfall of Atlantis (which also went by other names) brought about by descendants of the last living Edenites and their reliance on and blind trust in a massive technology.

Spirit world insight. In a computer contact,[11] our spirit friends told us that the Atlantean king's capital of Basilae was near the modern-day island of Helgoland. Atlantis was a marauding maritime culture with powerful technologies left over from the Edenite colonies. "The Project" was started 20,000 years ago (presumably by the superhuman castaways from Eden or their descendants), and it entered its final phase in Atlantis. Project Sothis involved a gateway through space-time which allowed people to communicate (and even to travel) through other dimensions to other worlds. It was hoped that Project Sothis would perfect the gateway to finer realms in order to salvage mankind on Earth. But alas, around 1220 BC, the advanced technologies of Atlantis got out of hand, resulting in a massive explosion, causing Atlantis to sink into the ocean and geologic upheavals to ravage Europe.

Hard evidence. German archaeologist Juergen Spanuth said that while in Egypt he deciphered hieroglyphics telling a story of a Lost Empire in the North. He reported that ancient artwork in the Egyptian temple of Medinet Habu showed Egyptian soldiers at war with sea people wearing horned helmets like those worn by the later Vikings. In 1953, soon after making his Egyptian discoveries, Spanuth traveled to the North Sea and found undersea ruins of what he believed to be Atlantis off the German island of Helgoland. Diving 13.7m (45 ft) his divers found red, black, and white rock walls.

Bones of Cro-Magnon people found in Europe are 30,000 years old and seem to be the first humans on Earth with the characteristic Caucasian appearance. The British National History Museum has recreations of various early humans that can be observed and compared on-line. (Go to *www.nhm.ac.uk/piclib/* then search on "cro-magnon".)

A series of earthquakes ravaged many Bronze Age cultures for a period of 1,200 years, according to evidence uncovered by various scientists—most notably the prominent French archaeologist *Claude-Frederic-Armand Schaeffer*. The major cataclysms occurred around 2300 BC, 1650 BC, and 1220 BC.

(I suspect that those cataclysms were the result of the death throes of Atlantis brought about by their technology.)

Soft evidence. The prophet Edgar Cayce, during one of his trance channeling sessions, reported that a powerful Atlantean technology called the Great Crystal, or Firestone, was housed in an oval building with a roll-back roof to allow light from the sky to charge the towering, six-sided crystal. The crystal itself had an adjustable capstone to regulate both the input and output of energy. The input included free-ranging energy from the sun, moon, and stars; from the air, land, and sea; and from sources beyond the electromagnetic spectrum (that is, subtle energies). These massive energies were stored below the capstone and beamed outward for various purposes.

Plato wrote that the Atlanteans built a palace surrounded by a stone wall, with towers and gates on the bridges beside the sea. "The stone which was used in the work," Plato said, "they quarried from underneath the center island.... One kind was white, another black, and a third red, and as they quarried, they hollowed out double docks, having roofs formed out of the native rock. Some of their buildings were simple, but in others they arranged the different stones, varying the color (white, black, and red) to please the eye, and to be a natural source of delight."

CHAPTER 8

Project Sothis

My hypothesis. A project called Sothis was begun thousands of years ago by some of the last living descendants of the Edenite castaways. The main aim of Project Sothis was to open portals to nonphysical dimensions. These would be portals not only for communication—allowing wisdom of the ages to stream into our world from the finer realms of spirit—but also for transportation. Humans would be able to leave the Earth through the portals, enter subtler universes, move instantly to other locations, and then re-enter the physical universe far away from where they started out. That way, "space travel" would not be a tedious matter of acceleration through the vast soup of space-time as envisioned by early science fiction writers (and even considered by most scientists today), but a quick and easy means of "blinking out" of one galactic neighborhood, entering a different dimension or parallel universe beyond time-space, then moving at the speed of thought before "blinking into" another galactic neighborhood in this physical universe.

The Project was not completed in Atlantis and was buried in the sands of time when the First Epoch came to an end. The hope seems to be that we today are fine and spiritually attuned enough to bring Project Sothis to fruition—that is, to open portals of other-worldly communication and transportation toward the goal of transforming this planet to a paradise.

I believe the Project has been a lot more challenging and grueling than expected. While the noble side of humanity resonates at a fine pitch, our savage side is almost incorrigible—a lusty, aggressive thing of legend, drama, and cinematic gore. The savage compulsions make it nearly impossible for us to resonate with the finer spiritual forces that would make Project Sothis possible.

So, much depends on us, at this juncture in human history, and on the decisions we make as individuals, as communities, as nations, and as a single living species on this Earth. Will the Project bring paradise to Earth, or will it be scrapped? Will we be able to forge a lasting relationship with Ethereal beings in the near future, as we nearly did with INIT, or will we plunge into a global dark age? That will depend on human choices in the near future, I believe.

What are the implications for us as physical beings, as social beings, and as spiritual beings? That's addressed in Part Two and Part Three.

Ethereal truth. In the spring of 1996 the Seven Ethereal beings first began making references to "Project Sothis" in their contacts to us. I had met with European researchers the previous year, and we had decided to forge an international association. We were told later that it was that earth-side decision to found INIT as an ethically motivated ITC network that compelled The Seven to call our initiative "Project

Sothis," suggesting it would be a continuation of the project in Atlantis.

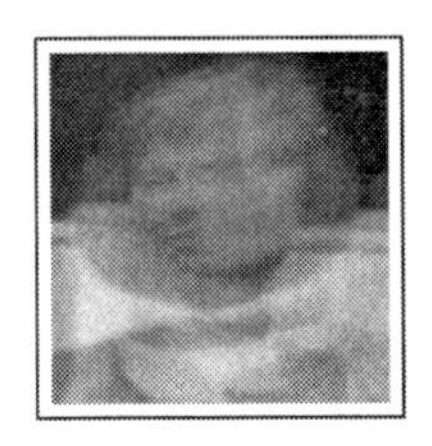

Spirit world insight. In the summer of 1996 our spirit friends began to provide some of the pieces to the puzzle of Project Sothis. They said it was modern humanity's oldest known project. By the time it got underway some 20,000 years ago, a space-time gateway had already been established in the area of modern-day Helgoland, the island off the German coast. People at that time were able to travel to other worlds through the space-time gateway, which they called a "dispassier point." The force of unity among colleagues on Earth is what made the Project possible then, and it's what can make the Project possible today, our spirit friends told us on many occasions. The contact field composed of everyone's attitudes and thoughts is reinforced by the unified cooperation of people who are involved in or concerned about the Project. A research group like INIT that can sustain harmony over time would be able to maintain the clear contact field necessary to resurrect Project Sothis today, allowing people not only to communicate with other dimensions and universes, but to visit them.

Hard evidence. Physicists today talk about "worm holes," the equivalent of our ancient ancestors' dispassier points. If you were to bend space-time like a sheet of paper so that you could stick a pencil through it in two places, that would be like a worm hole through which a person could move quickly and neatly from one place in space-time (such as Planet Earth) to another place—say, another planet in another galaxy.

CHAPTER 9

The Second Epoch

My hypothesis. Babylon marked the beginning of the Second Epoch of human civilization on Earth, and we could be approaching the end of that Epoch today. The First Epoch centered around Atlantis, and it had deteriorated into a dark age probably some seven to eleven thousand years ago, before the last flicker burned out in 1220 BC.

So the Golden Age of Atlantis—a time when parts of the Earth were like a paradise—was underway several thousand years before the energy experiments destroyed what was left of the civilization in northern Europe. At that time, the last remnants of the First Epoch were being washed away—literally—and the Second Epoch was taking root.

Civilization in the First Epoch included the fertile crescent formed by the Tigris and Euphrates rivers, known today as the cradle of civilization. A city called Shanidar was part of the empire, probably located around 400 miles north of modern-day Baghdad. A spiritual center, a temple called Sothis, was located along the banks of the Euphrates River on the same spot where Babylon would emerge thousands of years later. At that time long ago, the civilization fell into a terrible dark age at the hands of barbarian tribes, bringing an end to the First Epoch, and we today seem to be moving toward the next major dark age and the end of the Second Epoch.

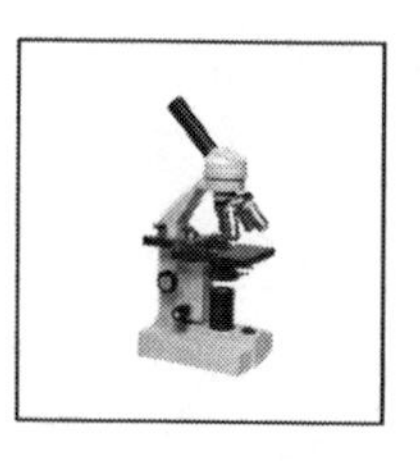

Hard evidence. Babylon was located just south of modern-day Baghdad along the banks of the Euphrates River around 2000 BC to 400 BC. Archaeological sites in and around the Shanidar Caves 400 miles north of Baghdad have produced evidence of farming around 10,000 BC. The caves themselves have yielded Neanderthal skeletons dating 60-80,000 years ago.

Ethereal truth. In the summer of 1998 our INIT research group received a contact from The Seven Ethereal beings through a computer at our ITC receiving station in Luxembourg. I publish below an English translation of the transcript in its entirety. The original contact arrived in German. It describes what unfolded on that final day of the First Epoch in an empire long ago, before its collapse into a dark age.

Alkbrat of Shanidar was the last one to enter the great temple hall. He walked bent over and was slow of step. His white hair and flowing beard gave him a patriarchal appearance. He leaned his curved golden staff against the table and took his place at the head of the table. The others in the room remained silent; they knew they had to wait until the venerable old man opened the assembly. It was the last time that they would meet like this in the Sothis Temple, for the barbarians were already at the outer walls of the city. Alkbrat raised both hands and began with a clear, firm voice, which, despite his age, sounded full and strong:

"Participants of this project, my brothers and sisters, the Golden Ages of our culture are at an end. Our civilization has become so indifferent that the innumerable dead, who die of starvation and disease, have become a common sight for you. They elicit no more than a sigh, or in the best case, a tinge of protest. The streets of our cities have become the residence of legions of the homeless; drugs rule the world; men murder their brothers on a scale and with a bestiality that has not been seen since the (previous) dark ages.

"At the same time, thousands of lesser acts of violence and abominations have become so customary that they hardly seem real. We have tried to avert the evil with our Project Sothis, but we have failed. We have made the mistake of reaching an understanding with the pupils of Nephtos, because we believed that as the caste of scientists, they would help us convince the people that the Gateway, the Space-Time Arch, was the last hope for our sick world. But we have erred. They have measured, minced and counted, and again, measured, minced and counted, and could not recognize the true meaning of our search, because they could not recognize the truth that stood behind our endeavors: the inseparability of the spiritual from the material world—the opening of the door between these worlds. So we lost precious time, and our adversaries used our weaknesses. Some of us have fallen away and let themselves be lured by passing fame and Mammon; others were intimidated and have lost their belief in the righteousness of our good cause.

We do not wish to bear them ill will, but rather pray that someday they will recognize their error. But we must be conscious that pride, covetousness, and arrogance are guilty of bringing our Project to ruin. Today we meet for the last time; tomorrow some of us will already no longer be among the living; others will be in flight. We know that our bodies, which we now inhabit, are only houses from which we will soon be moving. Where to and when we know not, but now we want to speak the last vow, and give our word, that at the given time, in another place, with another appearance, and under different names—some will be men and no longer women, and vice versa—we will no longer remember here and now, but when we meet again, we will know that we belong together. This very night, a Spirit of the Rainbow appeared to me and assured me that they will guide us further and lead us when mankind approaches the next dark age. This will be when voices speak from boxes and human beings move through light behind glass. Here on this spot on the river after many millennia, a city will arise that will be called Babylon, and the Second Epoch will begin."

Alkbrat arose. The golden threads that were interwoven through his deep blue garment shimmered in the glow of the candles. The others also arose and, extending their hands, they spoke the vow.

So it came to pass, and it is being communicated to you (INIT members), because you have found each other again, and this time have passed the first part of the test. Continue to watch out for false friends and know that the journey is a long one.

The Seven, 1998 July 3

In other contacts we were told that several of INIT's members and interested supporters had worked on The Project thousands of years ago, in other incarnations, and that is why they were now, in their current incarnations, especially drawn to ITC.

(In any case, as I mentioned earlier, I believe that after the collapse of Atlantis, the powers-that-be decided to close the case on the First Epoch, and they literally washed away much of that civilization with a major flood. The hypothesis is supported by ancient legends of a major flood, by surviving evidence of a great ancient civilization that extended throughout much of Europe and the Middle East, and by our spirit friends' testimony that Atlantis was a highly advanced, flourishing civilization.)

CHAPTER 10

Last Chance for Peace

IF YOU'VE REACHED THIS point after reading only my opening hypotheses in chapters 1-9—then it's time now to go back to Chapter 1 and read it all. I'll be here waiting....

The Ethereal beings told us in a computer contact on 1996 March 8:

We have often given you the real purpose of ITC contacts: Mankind at the end time should be led back to the principle. Light and darkness shall unite and form a whole again. What people experience now is not the actual beginning of the apocalypse, but only the first symptoms of it. Before opposites can be united, the strength of unity among ITC people must increase and come from a pure heart.

There are many of us alive today who share a purpose. We feel a pull to reverse humanity's downward spiral—to join with other world-servers and Light-workers to turn the tide, to help bring Project Sothis to fruition...to bring paradise to this world. But there is also a savage aspect of our nature that undermines our best efforts, so our destiny as a species rests most heavily upon the age-old battle raging within us.

It may be too late to avoid the coming dark age, or end time, and our progeny might have to wait to see if and when we humans will get another chance with a Third Epoch. But I'd like to believe we still have a chance this time. If we can make a few important choices in the coming years, we might—just might—be able to turn things around. There's still a chance to restore love to the world and to forge fruitful, durable relations with the light, ethereal realms of existence.

To bring paradise to Earth, we will definitely need divine intervention—the collaboration of angels—and I know from personal experience that that is possible, but only if we bring our noble side to the fore, acknowledge their existence, and request their help.

So, I herewith invite you to join the ranks of world-servers, not just in moving through this book, but in taking the next crucial step for our world. Include the angels in your hopes and dreams and prayers! If they see that enough of us are serious about forging Heaven on Earth, I know they'll step in to help.

PART TWO

Remaking the Noble Savage

I am not an animal!
Are so.
No way.
Way.
Gr-r-r-r.

THE TERM "NOBLE SAVAGE" was coined by English poet John Dryden in 1672 and used romantically to suggest that without the toxic effects of civilization, the unfettered human in his or her natural state would express a certain nobility. I use the term more dispassionately here, referring simply to the fact that we humans have two sides to our personality, thanks to evolution, and thanks also to the ancient cross-breeding between the superhuman Edenites and the primitive humans of Earth long, long ago. One part of us has been called our god-side, our spiritual side, our higher self, our divine self, or our good side. For now let's call it our noble side. The other part has been called our animal side, our carnal self, our material side, our bad side, our ego, or our dark side. Let's call it our savage side.

Our noble side sees the world as a beautiful but troubled place and compels us to be compassionate toward the less fortunate, to serve others and the world selflessly, to be genuinely concerned about those around us. Our savage side sees the world as a hostile place and compels us to do whatever we have to do to survive and flourish in it, such as taking what we need, protecting it ruthlessly, and defeating those who threaten us.

Stated in simplest terms, if we want to find peace and happiness in this life on Earth, the most effective way is by acknowledging the drama stirred up by our savage side and by making choices in our day-to-day lives that are inspired by and made in the best interests of our noble side. The time-proven religions have given humankind rich resources to help their members take the high moral road, including the Dharma Sutras of the Hindus, the Ten Commandments of Judeo-Christians, the Four Noble Truths of the Buddhists, and the Muslims' Sharia. By adhering to such guidelines, or simply by understanding our savage side and putting it in its place, we humans can get in the habit of making good choices that raise our spiritual vibration. Through our day-to-day decisions we find that crucial balance between trust and wariness. In any case, taking the high road is easier once we know who we really are.

CHAPTER 11

Who Are We, Really

WE TYPICALLY THINK OF ourselves as physical bodies moving around in this physical world. In truth we each have several body-minds—the physical one and some spiritual ones—that are all jumbled together in the same package.

At this point I need to add a new word to the vocabulary of the book: *in-beyond.* It's a hybrid between "beyond" (commonly used to refer to the other side, or the afterlife) and words such as in, inner, inside, and in-between. So, for example, from our conscious mind and physical body we can explore 1) the outer world (the physical world around us), 2) the inner world (the organs, tissues, cells, and molecules within us), and 3) the in-beyond world (the many heavens, hells, and states of consciousness inhabited by angels, spirits, ghosts, extraterrestrials, and other nonphysical or quasi-physical beings). Someday a new vocabulary will evolve to describe more accurately the multidimensional nature of reality, but for now "in-beyond" can at least get us through this book.

Our spirit bodies exist in those in-beyond worlds, and they can be compared to radio waves:

- they're normally imperceptible,
- they each have a unique frequency or vibration that keeps them distinct from each other,

- the different frequencies let them coexist independently in the same space, and
- we can learn to tune in to the various bodies separately with certain techniques and technologies.

In other words, we are made up of several bodies that are superimposed over each other. The physical body is the densest, and it's the one that our physical mind (the mind of the brain and five senses) is most familiar with. Our spirit bodies become subtler and finer as we move toward the central self—the true self—the soul.

That subtlest of bodies is the eternal flame—the source of inspiration for the Olympic torch—a small ray of undying light, a glowing fragment of the source that is often called "God." For centuries, mystics have called the heart "the seat of the soul," because typically that subtlest of bodies resides in the chest, glowing at the fastest and finest vibration imaginable, far beyond the perception of our senses. Hindu teachings describe the soul as a brilliant living light about the size of a thumb residing in the heart.

Between the soul (our true, eternal self) and the physical body (our here-and-now, temporary self) there are two spirit bodies—the astral body and the ethereal body. The astral body is in the form of a subtle human body, and the ethereal body is a formless being of light. The astral body resides in the astral realm, and the ethereal body resides in the ethereal realm.

Now, that's all true, according to my research, but it's simplistic. Saying that we have two spirit bodies is like saying there are two life forms on Earth—plants and animals. It's essentially true, but just as there many kinds of plants and animals on Earth, our spirit bodies manifest in many ways. For the sake of this book, though, let's keep it simple.

Let's just say that the ethereal body—our ethereal self, or higher self—is brilliant, timeless, and a source of tremendous inspiration and insight...once we know how to open up communication channels between it and our conscious mind. It resonates at a very fine and subtle vibration.

Our astral body can be a light, brilliant, blissful being when it flourishes on the noble human attitudes of love, trust, good will, and knowledge. Or it can be a dark and dismal being with sunken eyes and scruffy features when it's motivated by savage attitudes of fear, resentment, doubts, envy, and so on. In its blissful state it resonates at a fine vibration and resides in a paradise world, but in its dismal state it has a dense vibration and finds itself in dark surroundings. I call this version of the astral body the "dismal body."

During our lifetime on Earth we help shape our astral body by the attitudes we foster or fester in our day-to-day lives, and when we die, we settle into a spiritual realm appropriate to our astral body. There are many beautiful paradise worlds, and there are many dark, dismal worlds, and after we die our spirit moves automatically to the world of compatible vibration.

That's why religions and esoteric schools throughout history have encouraged people to make good, moral choices in life; moral living helps to raise the vibration of our astral body. That's the meaning of spiritual purification.

The diagram above simply lists four realms—physical, dismal, astral, and ethereal—but these are just arbitrary divisions. Each of those realms consists of many worlds and universes. Other people might refer to seven or eleven or 21 different levels or realms of spirit, but the numbers don't really matter, since they're all arbitrary anyway. It's just a question of how you prefer to divide up this infinitely complex omniverse. I like this particular model because it jives with how we normally think of the worlds of spirit nowadays. The dismal realm is where ghosts and lost souls get stuck for awhile, and when they go "to the light," it's the astral realm they go to. The astral realm is the paradise where most of us awaken after we die. The ethereal realms contain countless worlds beyond form and structure inhabited by angels and light beings.

Incidentally, most dreams are out-of-body experiences (OBEs) that we experience several times throughout the night while our physical body and mind are asleep, and our spiritual bodies and minds are awake and active.

CHAPTER 12

Why Do We Behave Like This

LOTS OF THINGS PULL us in different directions throughout our lives, compelling us to behave this way or that. Some of those things are inside of us, such as hormones and the nervous system. Other things are outside of us, such as our boss or parent or teacher, a stoplight, or an angry bee. And still other things that influence our behavior are in-beyond of us, such as guardian angels, ghosts, and departed loved ones, whose impressions we often receive in the form of voices in the head, urges, inspirations, and conscience.

Spiritual Influences

The "higher self" is a popular term used to denote our formless, ethereal body. It, too, resides in the in-beyond and influences us in powerful ways. In fact, Ethereal beings and our guardian angels frequently communicate with us through our higher self, which then tries to trickle-down the information to our conscious mind.

Loved ones sometimes stay attuned to us after they die and silently share loving thoughts and feelings with us, giving us a sense of warmth and contentment. They often "stop by" on special occasions—anniversaries, birthdays, and holidays.

Many brilliant, gifted men and women keep an interest in the affairs of Earth after they die. If we focus our energy on particular careers—music, art, or science, for example—we often attract these brilliant spirit friends into our lives, and their inspirations come to us in the form of, say, a new song or an important equation.

On the other hand, swindlers, murderers, and predators carry their troubles with them to dismal realms after they die, and they too can influence susceptible people on Earth. Techniques are offered in Part Four to protect ourselves and our families. For now let's just say it's good to keep your spirits up, so to speak!

The Brain and Its Reward Pathways

Our brains have evolved (and were probably engineered by Edenites as well) with reward pathways (also called pleasure centers) that are stimulated when we indulge in things that perpetuate our survival. Science today is busy learning amazing things about these reward pathways in our brains, which are stimulated by such activities as:

- eating,
- having sex,
- acquiring things, and
- making fight-or-flight decisions in stressful situations.

Dopamine is secreted into the brain at such times, and we feel waves of pleasure. It's a built-in reward system to compel us to survive. Scientists have determined that people tend to make choices in their day-to-day lives that activate the pleasure centers to get that shot of dopamine...a bit like lab rats pushing buttons over and over to get cheese.

Dopamine is a hormone that acts as a natural mood-altering drug, and we can become addicted to it. Anything we do that causes dopamine secretion is a potential addictive behavior pattern. Of all of the many dopamine behaviors, two in particular are generally considered to be healthy addictions—exercise and meditation. One strengthens the body. The other gives cohesion to the mind and spirit. In fact, twenty minutes of moderate exercise in the morning can set up a good dopamine level in the brain that can last most of the day. Twenty minutes of meditation can make you blissful and calm. More about that in Part Four.

The more that scientists study these reward pathways, the more triggers they find that activate them, and most of the behaviors that trigger dopamine are potentially harmful:

- Being aggressive, or even observing aggressive behavior, triggers the brain's reward centers, maybe because animals typically get aggressive to acquire things (mates, territory...) and to hold onto them—a definite asset for survival.[12]
- Pain in the body triggers the reward pathway to provide long-lasting relief.[13]
- Nicotine, caffeine, and recreational drugs have a huge effect on the reward pathways—essentially hijacking the circuits, overshadowing the dopamine, and making us crave more of the foreign substance...leading quickly to addiction.[14]

Dopamine in human brains has boiled over into the myriad dramas that make up life in societies everywhere—sports competitions, movies and books of crime and passion, gambling, romance, dining, shopping, criminal behavior, sadism, masochism, drug and alcohol abuse, sexual addiction and deviancy. Even the global stock market and international wars could be attributed in large part to our built-in penchant for drama, thanks to the reward pathways in our brains and the dopamine that triggers them. So it could be said with some truth that we have a built-in compulsion toward bad habits and destructive behavior, and it undermines our need for peace and collaboration in human affairs.

The reward pathway isn't our only built-in saboteur, though. At the back of the head is the brain's insular cortex, and it too compels us to make bad choices.[15] That small brain lobe gets activated when we get in a situation where our drug(s) of choice (alcohol, marijuana, nicotine, cocaine, etc.) are present. The insular cortex nudges us to partake in them. There are documented cases of heavy smokers damaging the insular cortex through surgery or brain trauma and completely losing the urge to smoke.

So between the reward pathway, the dopamine, and the insular cortex, it's as though we were programmed to be naughty long, long ago because we had to survive among nasty creatures in a ruthless environment, and even though our environment has mellowed out considerably, thanks to civilization, we haven't changed much biologically. As a result, we have a strong penchant for drama, and we build drama into every facet of our lives and every facet of society.

With that in mind, maybe it's time to do a little reverse engineering on ourselves to come out with a new product—a human being whose pleasure centers are activated when we make choices of good will and decency rather than brawling, boozing, drugging, gambling, gluttony, greed, and hot monkey sex. We'd certainly lead more peaceful lives, and we'd probably live longer too if we kept our savage side at bay by rewiring the reward pathways in our brains to bring out our noble side.

I'm sure some will argue that we may not live longer, but it'll certainly seem that way. ☺ Others will insist that our humanness has been designed this way for a reason (whether as part of God's plan or as the natural order of things or whatever), and we shouldn't tamper with it.

I'm convinced otherwise. If there's any hope for this troubled world, we humans need to start reinventing ourselves to be good. As civilization spreads, the world needs to become a more peaceful place. As society gets more crowded and wilderness keeps shrinking, we need to learn to rub elbows with each other without stepping on each other's toes. In short, we need to bring our noble side to the fore and move our savage side to the back burner to simmer.

There are a number of ways to do that. We can reinvent ourselves on a personal basis to some degree with the age-old, time-proven practices of prayer and meditation, through the refined art of body-energy healing, and by employing more modern technologies such as Hemi-Sync. In the future, science will help us control our savage side in much more effective and immediate ways—especially with brain implants (in the near future) and genetic engineering (in the more distant future).

I should note here that dopamine is just one of many hormones that tingle our savage side. There are others. At the risk of oversimplifying, testosterone can make men aggressive and sexually motivated, estrogen can make women moody and depressed, and progesterone can make women sexually frisky and men aggressive toward kids. Hormones rule the world, as they say, but dopamine is a good starting point when it comes to understanding biological reasons for our behavior.

Ultimately we each have the responsibility to find and commit to the noble path in our own way, getting to know the inner savage that makes us each unique and troublesome in one way or another.

And this may be the most important lesson that has restored sanity to millions of hopeless men and women: If it ever becomes too daunting a task to control our savage side—when things seem hopeless, especially in some sort of addictive behavior—we can simply admit that the situation is out of control, and we turn it over to a Higher Power. Once we do that, invisible forces with tremendous power and wisdom always move in instantly to begin restoring order and inner strength to our lives. Always. And as long as we have the will to take the noble path, despite occasional mistakes, relapses, and moments of weakness, the help will come.

CHAPTER 13

Working on Ourselves

Meditation and Prayer

Prayer is sometimes defined as talking to invisible superhuman powers (especially God), expressing our gratitude to them and asking their help. Meditation, then, is the art of listening to those powers. Both techniques can mobilize powerful forces in the finer realms of spirit to make miracles happen in our lives.

Prayer can be described as focused intention and a request for help. When we relax, focus our mind, set an intention, and state it in words, we send out streams of consciousness that mobilize powerful forces in light, lovely worlds. Our focused thoughts actually help shape subtle realities beyond our physical world. The realities we create and the forces we mobilize in those in-beyond realms then feed back onto our world to affect our lives. Prayer can help this process unfold in a most beautiful way when done from the heart with a clear, focused mind.

Throughout our lives we can always ask for help from the many Ethereal beings (angels) who provide guidance and protection to Earth and its inhabitants. As we align to those brilliant beings, we raise our spiritual vibration, which strengthens our noble side and weakens that savage voice inside that compels us to make bad choices.

We can pray directly to God, even though God is the Principle and as such cannot talk to individual beings (so the Ethereal beings told us). There are countless powerful intercessors who can help anyone, anytime.

Meditation is a time-proven technique for slowing down the brain with beneficial results. The brain operates at different speeds, from very slow during a deep sleep, to very fast when we are excited. Measured in cycles per second (cps), brain activity is often classified as follows:

BRAIN SPEED	MENTAL STATE	
	Nickname	Common Name
2-4 cps	Delta level	Deep sleep
4-7 cps	Theta level	Normal sleep
8-13 cps	Alpha level	Unconscious dream state
14-25 cps	Beta level	Awake
30-50 cps		Hysteria
50+ cps		Psychotic frenzy

Normally when the brain slows down to speeds less than 14 cycles per second, the conscious mind switches off and we begin to fall asleep. Physiologically speaking, meditation is simply the practiced skill of preserving a degree of conscious thought as the brain slows down; instead of entering a sleep state, we enter a meditative state, and a number of healing changes take place in our bodily routines, including deep, even breathing, slowed heart rate, reduced blood pressure, increased skin resistance, lower muscle tension, decreased metabolic rate, changes in the concentration of serum levels of some body fluids, and yes, a release of dopamine in the brain.

So, the trick with meditation is to slow the brain down without nodding off ... while staying somewhat alert. Usually all it takes is relaxing the body and clearing the mind, then the brain slows down by itself. These simple steps can be used:

- Get comfortable in a sitting or lying position and close your eyes.
- Starting at the toes and working slowly upward to the top of your head, concentrate on relaxing all your muscles.
- Clear your mind of random thoughts. This is sometimes easy, sometimes difficult. It may help to count slowly, silently backwards from 50 to 1 (or from 10 to 1 over and over), or to visualize a pleasant situation like a sunset, or to repeat a relaxing phrase such as, "slow, smooth, relaxed; slow, smooth, relaxed ..."

Benefits of meditating once or twice a day include diminished stress, a more relaxed demeanor, improved intuition and insight, more reliable "hunches," a higher proportion of correct decisions, better organized thought processes, a capacity to sense when others are troubled (and why), and much more, all the result of cleared communications between the conscious and unconscious mind.

In spiritual terms, meditation lets our mind leave the physical realm to travel in-beyond to finer realms flourishing with life richer and more real than this illusory Earth. Meditation takes us closer to the ultimate reality that is sometimes called the source, God, Allah, Yahweh, or Brahman. As we enter these finer realms, we are overtaken by bliss and inspiration.

Heart meditation is for me, at this point in my life, the ultimate spiritual practice. For centuries, mystics have called the heart the seat of the soul, and the soul an eternal spark of God. From my own experience, doing a heart meditation (moving my awareness from the head to the heart while slowing the brain) gets my conscious mind closer than otherwise possible to my true self—my soul. In 2006 I worked with Monroe Products to produce an audio "Hemi-Sync®" CD called Bridge to Paradise (see "Binaural Beats" below). It takes listeners through a heart meditation to a trip to a paradise world where many people awaken at the end of their earthly lives—a subtle, beautiful world beyond the physical—a world that Christians and Jews call Heaven, what Muslims call *Jannah*, what Hindus call *Pitraloka*, and what Spiritualists call The Summerland. The CD can help us get closer to our finest self.

Breathwork

One of the most debilitating human conditions comes in the form of locked up pain left over from emotional wounds suffered long ago—maybe from childhood, maybe from broken relationships, maybe from other lifetimes we have lived. The emotional pain can disable us in our interactions with other people, and it can pull us into addictions. One way to break free of the pain is breathwork, of which there are several forms. The technique that worked for me is based on Stanislav Grof's "Holotropic Breathwork," which involves hyperventilation.

I would lie on my back with feet flat on the floor and knees up, breathing as quickly and as deeply as I could. After about five minutes, the breathing would take on a life of its own, and in ten or twenty minutes intense, long-buried emotions would start to pour out. Rage and resentment from long-forgotten situations startled me at first, but the more I did this breathwork process, the more welcome they became, because very often when the process was finished I felt as though a ten-ton weight had been lifted off my mind and spirit! All my organs and tissues were completely relaxed. If I had trouble sleeping on a particular night, a session of breathwork would often let me fall into a deep slumber.

Tips from Tibet

The Dalai Lama, leader of Tibetan Buddhism, recommends several daily practices that quickly strengthen our noble side:

- Start the day with a five-minute reflection on the fact that we're all connected to each other and we all want the same thing—to be loved and to be happy.
- Spend another five minutes cherishing our self while breathing in, and cherishing others while breathing out.
- Throughout the day practice cherishing everyone we meet.
- Continue to do this despite our moods and the way others treat us.

Doing this daily can quickly polish up our noble side and etch a contagious smile on our face throughout the day.

I've developed these principles into two simple mantras and integrated them into a set of physical exercises. Described in Part Four, these "mantric exercises" can develop our mind-body-spirit connection and make us happy and congenial on a day-to-day basis.

Binaural Beats

At age 42, Robert Monroe started having spontaneous out of body experiences (OBEs) and had no idea what was going on. He could look down and see his sleeping body, and he could travel instantly to be with friends and neighbors. He could see them, but they couldn't see him. Sometimes he'd talk to them later and learn that they indeed had been doing at those times what he'd seen them doing during his unscheduled, unnoticed visits. After hundreds of such experiences, Monroe wrote a series of books about his in-beyond adventures. He developed an audio technology around *binaural beats*, patented it as *Hemi-Sync®*, and built a research center—*The Monroe Institute*—to further study it and to share its benefits with others. [16]

With binaural beats, the ears deliver pulsing stereo sounds to the left and right hemispheres of the brain, and the brain synchronizes itself to the rhythms. By slowing the pulsing rhythm, Monroe could move the listener's mind into higher states of consciousness, in which the body was asleep and the mind was awake. Essentially his technique could help people enter meditative states in minutes or hours, which some mystics would spend months or years to learn.

Today thousands of people from around the world have used Hemi-Sync and other, similar binaural techniques to alter their consciousness. They relax, put on stereo headphones, and listen to special recordings containing soft, pulsing binaural signals that move them to finer states of consciousness. Music and narrative in the recordings gently urge their minds to stay awake as their bodies sleep.

In the future scientists and engineers may be able to find

frequencies that have a direct influence on the reward pathways of human brains, and those frequencies, along with supporting music and narration, could help us re-program the system to resonate with love and good will.

In spiritual terms, the binaural beats technique is an effective means of releasing our spirit from our body, to let it venture among the in-beyond worlds of angels and ancestors while our body becomes idle. We all take these spiritual journeys naturally every night while we're dreaming; our spirit explores while our body sleeps. Unfortunately we hardly ever remember the excursions through in-beyond worlds. Binaural signals let us journey there at any time of day and remember the experience.

Focusing on What We Want

Whatever we obsess about, we attract into our life, for better or worse. If we focus on things we desire, we pull them toward us. Invisible forces begin opening doors and removing obstacles so that the object of our desire becomes attracted to us like a magnet.

If we obsess about things we dislike, they too are pulled into our life. The more we fight against something, the stronger it becomes, maybe to the point that we lose control. That's why fighting an addiction doesn't work. What we resist persists.

When we focus on what we really want in life and summon the help of a higher power, it comes our way.

Overcoming an addiction involves giving up control over the addiction to a higher power, then refocusing our thoughts and desires toward what we do want. And at the core we all want the same thing. We're all one, and we all want love and happiness. When we determine what makes our body-mind-spirit healthy and happy and allows our feelings of love to flourish, and when we focus on that while calling on a higher power, that's when the tight grip of an addiction begins to break down and we begin to see light at the end of the tunnel.

CHAPTER 14

Getting Help From Others

Body-Energy Work

Today in the USA there are growing ranks of health practitioners who use healing touch, Reiki, chi kung (or qigong) healing, network chiropractic, breathwork, flower essence remedies, high-grade essential oils, and many other energy therapies that reconnect us spiritually. All of these and other techniques, products, and healing ways tap into the energy bodies within us. Holistic health (integration of mind, body, emotions, and spirit) today includes healing work that stabilizes and grounds the physical body and mind while raising the vibration in the entire system. The body-energy work process helps us evolve to be more whole (holy) so that our conscious self connects to our finer or higher self. As we heal and release fear, anger, and harmful mental constructs and patterns, we eventually align with love and service to the best of our ability.

We live in glorious times when a vast array of healers, teachers, and evolved humans can help us as individuals and as a species to grow in love and understanding. More and more medical doctors are acknowledging the benefit and effectiveness of energy healing and other "alternative" health practices.

Brain Implants

So much has been happening in scientific labs in recent years that today we could be on the threshold of a complete transformation of human behavior. I can envision a day when people with electrodes in their brains will think thoughts of love and good will toward other people, and computers will record the neural activities that go on in their brains during those reflections of decency. Once the computer clearly recognizes those neural patterns, then other electrodes will be activated in the brains to stimulate dopamine along the reward pathways whenever they have those thoughts of love and good will.

This altruistic activation of natural dopamine along the reward pathways would override or replace the natural activation that's triggered by brawling, boozing, drugging, promiscuity, gambling, greed, and gluttony. Almost overnight these individuals would become some of the most stable, grounded, and loving people on Earth.

These experiments could begin with lab animals, and the technique, once perfected, could be used on a small group of human volunteers. Decency implants might first be deployed for people who are morbidly obese or desperately addicted to substances ranging from nicotine to heroine—men and women whose uncontrollable bad habits are killing them.

That day might not be far off. In April, 2004, the US FDA approved the first clinical trials of brain implants that would allow paralyzed people to produce motion by the power of their thoughts.[17] In 2006 a man paralyzed by knife injuries was using brain implants to play computer games, control a mechanical arm, and check his emails. Electrodes in his brain monitored neural activity that occurred when he tried to move a paralyzed arm or leg, then a computer was programmed to become familiar with those neural patterns, identify them, and translate them into useful activity of the equipment.[18] So as the man tried to move his arms and legs, he actually moved mechanical devices, with the help of the computer.

Brain implants have been a routine treatment for thousands of patients with Parkinson's disease for years. Lately they're being used in clinical trials for epilepsy, depression, cluster headaches, Tourette's syndrome, and obsessive-compulsive disorder. Rats have been given brain implants that stimulate their dopamine neurons when they exercise, compelling them to run treadmills and lift weights strenuously for hours.[19]

It's not hard to envision couch potatoes with high blood pressure becoming dedicated joggers with strong hearts in a few short months. All it would take is some well-placed electrodes in their brains.

Musicians, writers, artists, and other creative people (who often seem to be especially susceptible to dopamine disorders—especially substance abuse) could become some of the brightest lights in the world when brain implants compel them to pursue their creative life purposes and passions instead of being pulled into bad habits and addictions.

If the brain implant procedures are prohibitively expensive and not institutionally sponsored, wealthy patrons of the arts could foot the bill as a charitable donation, or they could consider it an investment, receiving a percentage of creative fruits of the electrode-facilitated inspirations. There are many possibilities.

Pharmaceutical Drugs

Many healthy people today wouldn't be alive without their miracle drugs. When a patient's inner voice works together with a doctor's expertise, amazing healings can occur with the help of pharmaceuticals. Diabetes, heart disease, glaucoma, and AIDS are just a few of the syndromes that can be ameliorated with prescription drugs.

On the other hand, pharmaceutical drugs sometimes take a "shotgun" approach to treatment, causing more problems than they fix, especially in the area of psychiatry and behavior modification.

If you're making health decisions for yourself or your children, unless the drugs are an urgent matter of life and death (in which case heed the physician), the best advice is to be attuned to your higher self through inner work (especially meditation and prayer), learn everything you can about the drug and treatment possibilities recommended by a physician, then let your inner voice work with the physician on the prescription. Except in crisis situations, it's best to begin with the most natural of therapies before progressing to the most intrusive.

Shaping Our Genetic Destiny

Many of the genetic adjustments performed by the Edenites long ago will surely be redone someday, adapting humanity for a civilized paradise world rather than a wild, ruthless one. The reward pathway in the brain will be "rewired" to be activated when we choose to behave in ways that foster harmony among people. How exactly will that happen? Here's one possibility:

The genetic materials (DNA) in our body cells are made up of four alkaloid molecules abbreviated as A, T, G, and C. Genetic coding (how those four molecules are arranged in a long string) determines much of who we are. Any particular segment of code is called an "allele," and behavioral scientists have found something they call the "A1 allele" that is in about 25 percent of the people in the world. People with that code sequence in their DNA have fewer dopamine receptors, and so they exhibit more compulsive and addictive behaviors. It's real hard for people to quit smoking, for example, if they have the A1 allele in their genes.[20]

Since addiction destroys many relationships and ruins many lives, adjusting this A1 allele may be one of the first genetic alterations we'll want to make on ourselves, once a safe, effective technique has been developed. Then, on to other trouble spots in the human make-up. We know that certain hormones such as cortisone and certain brain regions such as the hypothalamus are associated with stress, aggression and violence,[21] and with a bit of tweaking there we might become more patient, relaxed, and generally peaceful.

New genetically engineered human beings might be our ultimate hope, dream, and destiny, but they'll have to wait until genetic engineering comes of age. At present it's a very mixed blessing.

On one hand there have been some great bioengineering success stories. Case in point: The ringspot virus was devastating Hawaii's papaya crop for years until the winter of 2006, when researchers from the University of Hawaii, Cornell, and the US Department of Agriculture finished engineering and rigorously testing a virus-resistant papaya. The following year Hawaii's papaya crop increased by 53 percent in October alone. Promising genetic engineering experiments have also been conducted to produce heartier fish, to cure autism in mice, to use bacteria to mine precious metals in environmentally friendly ways, to produce vaccines against various viral diseases, to make advances toward permanent cures for cancer and hereditary diseases such as sickle-cell anemia, and to develop blood-clotting agents, insulin and human growth hormones.

But for every genetic engineering success story there is also a horror story. Several hundred people died in Spain in 1983 after consuming a genetically engineered rapeseed oil, which was sold in stores because it was proven not to be toxic to laboratory rats. It hadn't been tested on humans. A few years later some 1,500 Americans were permanently disabled and 30 died after consuming a new brand of the popular over-the-counter dietary supplement L-Tryptophan. The genetically engineered bacteria used to produce the new brand of L-Tryptophan was contaminated during genetic alteration, and it was sold to unsuspecting consumers with no labels identifying it as a genetically engineered product. Since then, experts have issued plenty of warnings.[22]

I'm sure it'll be some years before we have the knowledge and proven experience to conduct safe, effective genetic engineering experiments on humans... but that day is certainly approaching. It may be the ultimate "fix" for humanity.

Meanwhile, personal development and personal responsibility allow people to develop their noble side. The tools of science and technology will provide more possibilities in the future but will never be able to replace or override our free will.

CHAPTER 15

Remaking Summary

IN SPIRITUAL TERMS, WHEN our thoughts, words, and actions reflect love and good will, our spirit vibrates at a fine rate, resonating with paradise worlds and Ethereal realms. As a result, we attract into our lives loving spiritual influences that are supportive, manifesting wonderful synchronicities and miracles in our lives. By reinventing ourselves through our own practical techniques or with the help of the experts (both in this world and in-beyond), we could each enjoy that inspired condition for the rest of our lives.

On the other hand, when we harbor thoughts of revenge (we are unable to forgive) or feelings of lust (we can't, or won't get that person out of our mind sexually), or if we're gripped by fear or resentment or greed, our spirit takes on a dense vibration that resonates with dismal, troubled spirit worlds, and we attract spiritual influences that stir up our dark feelings even more.

We each have to take responsibility for our thoughts, words, and actions. By making decisions that bring out our noble side or our savage side, we're altering ourselves spiritually in one way or another—raising or lowering our vibration. So by the phrase "reinventing ourselves" I'm referring to the things we choose to do to lighten up our spiritual vibration.

The whole subject of reinventing ourselves could stir up a lot of drama. Many noble-savage purists would rather see humanity struggle with its savage side (even if it means beating itself to death with weapons of mass destruction) than to surrender to a kinder, more decent future.

I prefer to see humankind rise to a new level in the coming years. There are things we can do now to foster our noble side and to start moving slowly into subtler vibration, and there are techniques that science will perfect in the future to help us to "lighten up" quickly and definitively.

Life on Earth is a rare experience; light and loving thought-forms are mixed together in society with dark and dense, troubled thought-forms. In the worlds of spirit the two types of thought-forms don't mix, just as radio waves of your AM rock station don't mix with your FM public broadcasting station. Different channels, different frequencies. Fine and crude don't mix in-beyond.

Throughout our lifetime on Earth we're a mixture of positive and negative emotions, light and dark attitudes. The light thought-forms stream into light spirit worlds where good will is the driving force of life. The dark thought-forms stream into dismal, troubled realms of spiritual existence of dense vibration. When we die, our astral body has developed a certain home frequency or vibration, determined by our thoughts and attitudes, and that spiritual vibration acts as a homing signal to carry us to the world where we resonate. Most people are carried by their noble side to paradise worlds in the astral realm after they die. For others, if the savage side has prevailed during lifetime, it carries them to the dismal realm after they die, where they remain stuck in troubled thought-forms for awhile—until they can resolve their inner turmoil and move smoothly "to the Light."

That's why our Ethereal friends used ITC systems to warn us of destructive spiritual influences and to reassure us that help is always available:

You already know that also pharisees, ghouls, swindlers, thieves—yes, even murderers—have their interested supporters here among the dead.... Even if we cannot avoid the plague (of such destructive spiritual influences that affect people on Earth), we can control the gravity.

The savage human condition on Earth is taken into account in the big scheme of things in-beyond. It's well-known "over there" that the noble human spirit is dragged down by a lifetime on Earth, so light beings are always trying to inspire us, and they're often there after we die to sweep us away to a beautiful in-beyond world, where our spirit body is rejuvenated to a fine vibration. While alive, we can often feel the presence of these supportive beings during our dreams, daydreams, and meditations, especially if we're in the habit of praying and meditating on a regular basis. And when we die, which is often a painful, troubling experience, beings from finer spiritual realms are usually present to help us bypass the dark, dismal realms and go straight to paradise, a world that vibrates at such a fine and subtle rate that to our spirit, fresh out of a dense Earth experience, it appears as a bright and beautiful but nearly blinding light. And it takes some getting used to. So when we arrive in this paradise world we usually enter a period of deep sleep lasting about six weeks of earth time, during which our old astral body grows in reverse to the prime of life and the peak of health, and our thoughts and feelings stabilize with a predominant sense of love, trust, and good will toward everyone and everything. In these worlds of light we are driven by a compulsion to serve others while partaking in the joys of paradise living.

Meanwhile, here on Earth we can get a pulse on the spiritual state of humanity by observing the subjects being talked about and enjoyed in people's spare time. Today we can observe the mass media, especially movies. Hollywood has been leading the way into the coming dark cycle with its rapid and widespread proliferation of lust, gore, and terror throughout the world. There are many beautiful, inspiring, uplifting movies produced as well, but they usually don't get the attention and box-office receipts of the more savage films. That suggests that our savage side is tipping the scales at the moment.

Speaking of which, many creative people would like to express their most beautiful inspirations in a way that the public will embrace. How to touch the world with their love or intellect through their writing, music, art, and cinematography? As most of us learn quickly in writing and the arts, we sometimes have to figure out a way to wrap our ideas and inspirations up in drama, because drama is what stirs people. Drama is essentially conflict between the noble and the savage—whether the conflict rages between two individuals or two groups, or within the mind of a character. It's all about conflict—drama.

If we were living in a noble world where conflict is rare, the drama would be unnecessary, even repulsive to many. Just as if we were living in a savage world there'd be no place for the beautiful inspirations because everyone would be too busy dealing with their conflicts. But we're living in a noble savage world, so popular writers and artists who want to share their inspirations with lots of people, usually learn how to wrap them up in a compelling drama to reach the masses.

So what's the moral of the story? The plot? Well, it's this: If you want to attract loving guidance and protection into your life, the simplest way has always been prayer. Simply talk from your heart to the finer powers through prayer and ask for help, and those wonderful beings will be at your side instantly. Attracting supportive spiritual influences is easy. Just pray. And to *keep* those fine influences in your life, meditate frequently to raise your spiritual vibration. That enables them to be with you more regularly.

Likewise, attracting dark and troubled forces into your life is easy. It doesn't require any special incantations or rituals. Just let your mind be filled with fear or animosity or lust or any of the other savage emotions, and troubled spirits will be at your side instantly to stir up your life in unpleasant ways.

So we have a simple choice in life as to what sort of spiritual influences we want to attract to give us support. It could be angels or brilliant scientists or artists or musicians in spirit, or it could be pharisees, ghouls, swindlers, thieves, yes, even murderers. It depends mostly on what we choose to focus on in our thoughts and attitudes.

In the near future, implants could be posted along the reward pathways of the brain to stimulate dopamine when we behave in ways reflecting love and good will. Or brain implants could be posted in the insular cortex to neutralize that area of the brain that compels us to partake in our addictions. As science learns more and more about the brain and its impact on our noble savage behavior, there'll be more and more possibilities for these implanted electrodes.

Meanwhile, let's take a look at the big picture—rethinking society.

PART THREE

Rethinking Society

OUR INIT GROUP RECEIVED the following message from the Seven Ethereal beings through the computer of our Luxembourg members on April 3, 1996:

In perceiving their environment, mammals, like humans, have the ability to evaluate their surroundings and behave accordingly. Though they are humanity's fellow creatures and inhabit the same living space, humans behave as if the world is their environment alone and everything else is only for their use.

*This anthropocentric world picture is a totally false self evaluation. Man is not the measure of all things as you often arrogantly assume. Humans are one of a million species on the tree of life. All the animals, plants and the elements of nature are part of the world around you. By living as though the rest of the world is only for your benefit, you miss the purpose of your existence.**

Pope Pius IX in his days opposed the founding of an Italian Society for the Prevention of Cruelty to Animals. He considered any obligations of men toward animals a theological error. After passing over in 1878 he was taught differently. Since then he has been cleaning

* I believe they are suggesting that that purpose is to be caretakers, not manipulators, of the Earth.

stables. We are told he is doing well and will soon be taking an active part in the nursing of animals. We are now waiting for some of his colleagues who issued a directive during the German Conference of Bishops in 1980 in which they underlined that human life takes precedence, and therefore medical tests on animals should be approved. Elsewhere, intelligent people like Rene Descartes turned animals into objects of human research curiosity...

Our noble side wouldn't tolerate our abuse of the planet and its inhabitants, but it seems to have equal say with our savage side, which can be especially brutal when we behave not as individuals but as societies. If we wish to fix the Earth, we'll first have to reassess ourselves as social animals with a savage side, and I believe we can do that best when we put humankind in perspective, as part of nature rather than apart from nature.

CHAPTER 16

Society, One of Five Living Systems

THERE ARE COUNTLESS VARIETIES of living systems in our world that can be classified in various ways. I like the following way, which involves just five basic groups:

Biosystems are the independent plants and animals we're all familiar with (birds, trees, people, cats, frogs...), as well as bacteria, viruses, and other organisms of all sizes. Things are well-organized inside a biosystem but more or less chaotic outside, depending on whether it inhabits an ecosystem, a social system, or an *ordisystem*.*

Bio-subsystems are the inner parts of biosystems, such as a heart within a person, or a heart cell within a heart, or tiny organelles within a heart cell. Life is very well-organized both inside and outside a bio-subsystem, which can't survive on its own and needs the larger system around it (the host system) to satisfy its needs.

Ecosystems (forests, oceans, jungles, savannahs...) are the wild places whose members (biosystems, social systems, and ordisystems) fight and kill each other for nourishment, territory,

* "Ordisystem" is another word I've had to create for this book. It comes from *ordi*, Latin for order. I tried to use the term "superorganism," but it's too general; it includes aspen groves and coral reefs, for example, and doesn't include the idea of protective enclosures (the outer membrane of a bee hive, the clay mound of a termite colony, or the tunnel walls of an ant colony) that protect an orderly system (ordisystem) from the chaos of the surrounding ecosystem.

and defense. Ecosystems are rife with conflict and disorder inside and out. Murder and mayhem are a way of life.

Ordisystems (honeybee hives, ant colonies, and termite colonies, for example) are tightly knit communities of biosystems living together compatibly within a protective enclosure, with the clear understanding that the needs of the community outweigh the needs of individual members.

Social systems are human groups ranging in size from families and friendships to nations and religions. Social systems aren't as tightly knit as biosystems or ordisystems, in which the needs of the group clearly outweigh the needs of individual members. Throughout history humankind has struggled to find a balance between the needs of individual human beings to be free and the needs of their groups to be stable and peaceful. This traces back to our noble-savage cross-breeding. Our collective nature (noble side) urges us to collaborate with and to serve and to trust others selflessly and honestly, while our individual nature (savage side) demands independence, harbors suspicions toward the motives of others, and bristles when someone tries to dominate us or take advantage of us. That inner conflict has boiled over into tensions and squabbles in nearly all human groups, and it has become the basis of the human drama, stirring up conflict in relationships ranging in size from marriages to international alliances.

If we could step back and observe all the life forms on Earth, we'd see that they don't all fit neatly into these five groupings. Some seem to be hybrids. Or, stated differently, the five groupings don't have a solid line between them; they sort of blend together as in the following table, in which white is at the orderly (noble) end of the spectrum, and dark is at the chaotic (savage) end.

Types of Life Forms	Examples and descriptions
Bio-Subsystems	Organs, body cells, and organelles are orderly inside, as well as outside in the cozy and complex "flesh-and-blood" world around them.
(hybrid example)	E. coli bacteria in the human gut are parasites (biosystems), but they behave like natural parts of us (bio-subsystems), helping us to digest the food we eat.
Biosystems	People, trees, cats, bees, and bacteria are orderly inside, but more or less chaotic outside in their surrounding ecosystems, social systems and ordisystems.
(hybrid example)	The Portuguese man o' war looks like a jellyfish (biosystem), but it's actually a colony (ordisystem) composed of four kinds of specialized polyps living together tightly within the confines of the organism. The polyps can't survive on their own; one polyp digests food for the colony, another procreates, and so on. The colony is so well-integrated that it behaves like a crude biosystem, floating with the current (unable to swim), but stinging and eating fish that swim into its tentacles.
Ordisystems	Bee hives and ant colonies are organized inside, but not outside in the surrounding ecosystem.
(hybrid examples)	Military boot camp, formal meetings, religious ceremonies, some communist-totalitarian societies, and other social systems with rituals, tight regulation, and specialized roles are sometimes so regimented that they're compared to insect colonies. An ideal example witnessed around the world recently was the awe-inspiring opening ceremonies of the 2008 Olympics in Beijing.
Social Systems	Families, companies, and nations are typical social systems—subject not only to their members' noble side (wisdom, empathy, honesty, trust...), but also to their fears, envy, resentments, and other savage moods, which stir conflicts and tensions within the group as well as with other social systems, so the typical social system rarely becomes as orderly as a biosystem or ordisystem.
(hybrid examples)	Run-down neighborhoods of gangs, drug dealers, pawn shops, porn shops, and liquor stores inspire the phrase, "It's a jungle out there," because of the violence, desperation, and predation among the people.
Ecosystems	Jungles, forests, coral reefs, savannahs.... Chaos and conflict are the rule, as life forms in the ecosystem kill and eat each other to survive.

So this is one easy way to classify the myriad living systems on Earth, and it lends itself well to our understanding of our own noble-savage qualities within the larger framework of nature on Earth.

Basic Building Blocks

It helps to understand living systems as being composed of basic building blocks. Though it's a simple concept, it's an important one in rethinking society, so kindly bookmark this in your mental notes:

- A biosystem example: The basic building blocks of a human being are body cells (bone cells, muscle cells, blood cells, nerve cells...) and molecules (hormones, enzymes, DNA...). The body cells work together and use the molecules to keep the complete system alive and healthy.
- An ordisystem example: The basic building blocks of a honeybee hive are the bees themselves and their products (honey, royal jelly, honeycombs...). The bees work together and use the products to keep the colony alive and healthy.
- A social system example: The basic building blocks of a nation are people and products (houses, clothes, foodstuffs, cattle, highways, pets, computers, TVs, ships, stores, farms, factories...). The people work together and use the products to keep the nation alive and healthy.
- Ecosystem example: The basic building blocks of a forest are biosystems (trees, squirrels, wolves...), ordisystems (ant colonies...), and social systems (forest homes and communities...), plus the products those systems need... which often include each other. Hence the tendency of

members of an ecosystem to fight, subordinate, and kill each other to survive.

Moving deeper into the structure of life, the basic building blocks of a living cell are organelles (or maybe some smaller living units composing the organelles?) and molecules that the organelles use to keep the cell alive and healthy. As we delve deeper and deeper into life's structure, into the atoms and subatomic particles and deeper still, we eventually reach a point at which matter fades away into energy—living waves... vibrations...consciousness. Consciousness seems to be the basic stuff of all life everywhere. And at the deepest level of consciousness, far beyond the realm of structure, everything is alive. Everything is one. That's where we find "The Truth," ultimate reality. What we find in this book is a tiny subset of that truth, bent and shaped to fit a miniscule and odd little part of the ultimate reality—that is, physical life in general, and humankind in particular. But I'm getting ahead of myself. Let's return to the basic qualities of living systems.

Feeding the System

Living systems must absorb part of their environment to satisfy their structural and energy needs inside. This is another simple concept that's crucial in rethinking society in a more natural light, so, again, please make a mental note:

- Biosystems: People and trees, as well as lions and rabbits and insects, eat food, drink water, and breathe air. These raw materials are ingested and used to satisfy the material and energy needs inside the biosystem.

- Ordisystems: Honeybee hives consume nectar from flowers, which is used inside the colony to make honey.
- Social systems: Nations consume natural resources (metals, timber, oil, ocean fish, water supplies, minerals in farmland, sunlight, wind power....). These raw materials are ingested and used to satisfy the material and energy needs inside the nation—that is, some of the resources are broken down into pieces that become part of the products and people in the nation, and some resources are converted to energy that gives motion, heat, light, and sound to the people and products.

So natural resources are the food of a society. Very simple, but very important. The following table helps put things in perspective so far:

Types of living systems	Basic building blocks of the system	Nourishment from outside the system
Bio-subsystem	Bio-subsystems and molecules	Nutrients provided by the host system
Biosystem	Bio-subsystems and molecules	Foodstuffs—groceries eaten by people, meat eaten by lions, grass eaten by rabbits, organic materials absorbed by trees through their roots...
Ordisystem	Colonial biosystems (such as bees or ants) and their products	"Food" (nectar, leaves...)
Social system	People and products	Natural resources
Ecosystem	Biosystems and ordisystems and social systems, and the products these systems use	Sunshine, rain, air, soil...

So we humans in our social systems are like cells in the human body, only different. We're also like wild animals in an ecosystem, only different. And we're like bees in a bee colony, only different. The differences are due mainly to our noble-savage cross-breeding long, long ago. Our noble side, inherited from the gods, urges us to serve and to be served by others in love, trust, and good will. Our savage side grew out of the wild terrestrial ecosystem and compels us to be as suspicious, fearful, greedy, and aggressive as necessary to survive in unpredictable, often hostile surroundings.

So the first basic principle for society's well-being is this:

Principle One

People and their societies are of nature, not above it.

CHAPTER 17

Life's Nested Structure

IF WE WERE LOOKING down on the Earth from space* as though through a telescope, we would not see political borders or country names. Instead we'd see cities and farms, highways and other organs and tissues of the human species speckling the global landscape like patches of mold on an old peach. If we could focus in only on social systems to determine what exactly they're doing, we'd see large networks and industries for:

- communication (Internet, phone networks, radio and TV signals...),
- transportation (highways, railways, airline routes...), and
- manufacturing (transnational corporations with their farms and factories), among other things.

Each major subsystem in humanity, in turn, is composed of smaller organizations, which are composed of still smaller groups, which are composed of people, the basic building blocks of society.

* Anyone with a computer attached to the Internet can enjoy that bird's-eye view of our world. By clicking on www.earth.google.com, downloading the Google Earth program, and running it, you can see the entire planet or zoom in as close as a neighborhood to look at individual homes and businesses. (I saw our family car in front of our house!) By default, the Google-Earth photos are superimposed by city names, borders, and other artwork, but those can be removed for an uncluttered view of Earth by clicking on "Primary Database" along the edge of the screen.

But it doesn't end there, of course. Life on Earth is a chain of nested systems (that is, systems within systems within systems...). Looking inside ourself, we contain several large biological subsystems for:

- communication (nervous system),
- transportation (circulatory system), and

manufacturing (digestive system), among other things.

- Each major subsystem in our body, in turn, is composed of organs and tissues, which are composed of tiny cells... downward... inward....

Looking outside our self, back up the chain of nested systems, we are part of several social groups, including perhaps family, company, church (or mosque or temple or synagogue), clubs, professional organizations, and friendships. These in turn may compose larger and larger social groups. Family, for example, is part of a neighborhood, which might be part of a city, which is part of a state or province, which is part of a nation-state, which is a member of international alliances... upward... outward....

So the human being, like any other living system on Earth, is one link in a nested chain of living systems, which for us humans include body cell within organ within person within family within city within nation.... In short, world society is a rich tapestry of nested, overlapping systems.

A cohesive life structure depends on the commitment of each system at every level to serve as a healthy, reliable link between its inner world and its outer world. For example, you and I would each make choices that strengthen, stabilize, and integrate our body, mind, and spirit (internal concerns) and the groups we belong to—family, friendships, company, community, etc.

(external concerns). Our inner choices might involve exercise, meditation, and nutrition. Our outer choices might involve our family and the company where we work—the things we can do to give strength and stability to those groups through our knowledge, skills, and good will toward others.

Another example: A nation does its part in the structure of life by making good choices pertaining to 1) internal concerns (education, transportation, communication, science, industry, health care...) and 2) external concerns (using resources wisely, caring for the surrounding ecosystems, sustaining peace and productive alliances with other nations, playing a responsible role in international groups and the United Nations...).

So this is Principle #2 in society's well-being (it's actually a principle for the well-being of all life on Earth, including societies):

Principle Two

Life flourishes when each living system serves as a healthy, reliable link between the systems inside it (internal concerns) and the systems of which it is part (external concerns).

The highest link in the nested chain of society is the United Nations. The UN is a beautiful idea whose time has finally come. We've spent a half century letting it test and strengthen its wings, and now it's time to fly. We need a force of order and stability at the highest level of humanity—a force that can help to sustain and coordinate peace and equity among nations, religions, transnational corporations, global networks of communication and transportation and other macro systems. It's time to give the UN the authority of a world government. Today it exists largely as a gathering of autonomous nations exerting their influence on each other with a degree of protocol. Its mission thus far has been to foster peace and knowledge in the world... but with its hands tied. It's time to free the hands. More on that in the next chapter.

Addiction, the Disabler

Addiction can destroy the coherence in the nested structure of life. Most of us have at least a general idea of what addiction is among us humans, but I believe a similar unhealthy condition can affect life on Earth at any level. Cancer could be regarded as addiction at the cellular level. The petroleum crisis today could be regarded as addiction at the societal level (a subject of the next chapter). So, let's take a closer look at addiction, a major obstacle to a coherent nested structure.

Addiction is any relationship to a substance or an activity that makes life continually worse, yet the person (or living system) continues regardless. For us humans, addiction is often the result of emotional wounds received during childhood. When children's deep-seated needs for love or acceptance or appreciation are not met, they tend to bring secrecy and mistrust along with them into adulthood, and they develop behaviors or seek out self-medicating drugs and alcohol that provide momentary comfort but make life continually worse in the long run. Until they heal those early wounds, as through therapy or spiritual development, they will never be able to enjoy happy, trusting relationships with others. They remain uneasy in friendships and at work. They become dependent on their drugs and behaviors of choice, and remain trapped in the vicious circle of their addiction, thinking that their behavior will improve their mood when actually, except for the initial, short-lived highs, their mood grows continually darker.

When we support and assist family members or close friends trapped in on-going struggles with drug abuse, eating disorders, alcoholism, workaholism, sexual misconduct, gambling, compulsive shopping, or another form of addiction, we are assuming they will eventually free themselves from their problem. If they fail to do so, our continued support eventually begins to enable their addiction and self-destructive tendencies. We begin to make the problem worse. What they really need is beyond our ability to provide. The solution is beyond our world and accessible only to them. Substance abuse professionals and spiritual counselors can help only so far. The solution is in-beyond, in the acceptance of a Higher Power.

Overcoming Addiction

The most effective and widespread addiction recovery program ever devised is Alcoholics Anonymous, and the first two steps in its famous 12-step process are: 1) We admit we're powerless over alcohol, and our lives are unmanageable. 2) A power greater than ourselves can restore sanity.

From my own experience I know that there is no authority on Earth that can provide the wisdom, love, and transformative potential of an effective "higher power" to free us from addiction. We have to look beyond our world, and that's not easy for many of us to do. We're caught up in what we perceive through our five senses—the material world. We've separated our conscious selves from our living spiritual selves. That separation has created a fear of death, which spins off to create many of our other fears, and to stir up our addictive compulsions. That separation of mind and spirit also boils over into society, compelling nations to develop fearsome armies and arsenals. As our Ethereal friends told us:

Fear of death is one of the most distressing concepts of human culture. It is based on the conscious belief that your bodily existence offers life and security, which it never wants to lose. Fear of death therefore is evidence of the mind having lost its roots. It shows a spiritual being who has far removed itself from its higher self. You owe this mentality largely to an intellectual and scientific way of thinking. It wants all thoughts reduced to a comprehensible level of material existence. Heaven is in man and those who have heaven within themselves go to heaven. Heaven is in all those who recognize what is of God and let themselves be guided by the Divine. The priority and basic concern of every religion has always been the acknowledgement of God!

Meanwhile, movements are underway in some parts of the world to restore the spiritual fabric of entire communities. In the USA, dozens of organizations can be found online listing *intentional communities* and offering *community-building workshops*. In India there is the *Swadhyaya* movement, introduced to me by Majid Rahnema, retired Iranian ambassador and grassroots philosopher, who spent a lot of time in India, getting to know the people.[23]

The *Swadhyaya* (*swa* = self, *adhyaya* = study) movement began in India in the 1950s and grew from 19 members to several million all around the world, all devoted to self-realization (or connection to higher spirit) and divine brotherhood. *Swadhyaya* communities are devoted to wholesomeness. Economic betterment is not a goal of the movement, but those communities that embrace *Swadhyaya* are not just more wholesome, but more economically developed, cleaner, and more efficient…and the people are enthusiastic and mutually respectful. Money and property within the *Swadhyaya* community is regarded as impersonal wealth. It is earned by the members but belongs to God, and it is distributed discreetly and with grace to members where needed.

The movement is based on a lack of preaching or recipes for a better life. The only requirement of members in their dealings with non-members is to make it clear that the love and respect they show other people are gifts of the soul. When they visit people who are in need and who are outside the community, the members do not offer material support and certainly don't preach to them. All they offer is self-empowerment. Ancient wisdom suggests that it's okay to give apples to hungry people, but it's far better in the long run to teach them to grow and cultivate orchards. The nature of *Swadhyaya* is to cultivate the spirit. Coastal communities of smugglers, gamblers, and thieves embraced *Swadhyaya* and evolved into cohesive fishing communities.

Self-realization—exploring the in-beyond for that Higher Power through such practices as meditation and prayer—can help alleviate a society's addictions through its citizens' personal development. It can bring strength and coherence to life's nested structure as it manifests in society.

While *Swadhyaya* works well in India, it's probably not ideal for everyone and every culture. As I mentioned earlier, lots of good ideas can be found on-line under such key words as *intentional community* and *community-building workshops*.

CHAPTER 18

Who Decides What

REGULATION HAS A BAD reputation in some circles, mistaken as it often is for an urge to dominate, or a thirst for power, or endless spools of red tape. In fact, when done right, regulation is a vital, healthy process that promotes order at all levels of life on Earth. When regulation is absent (as in ecosystems) or when it's done wrong in society (when guided by our savage side), intimidation, bullying, domination, and predation take over, stirring up tensions and conflicts. In this chapter we'll look at the *right* way to regulate!

The Nature of Good Regulation

There are probably as many ways to describe or define regulation as there are names for it (management, leadership, administration, governance, ordinance, superintendance, guidance, direction…). If we tossed them all into a pot and boiled them down, we would probably wind up with two basic ingredients of sensible regulation — monitoring activities and making changes when necessary.

Monitoring doesn't have to be a constant vigil.

- We might monitor environmental degradation, weather patterns and the growth of civilization on the planet's surface with an occasional series of photographs from a satellite orbiting the earth.

- Monitoring a child at play might require an occasional glance.
- Monitoring employees in a company might involve an occasional report by each employee on the status of his or her projects.

Making changes is easy, but making the appropriate changes at an appropriate time is a bit more challenging:

- When to restrain the rapid growth of societies and industries to protect the oceans, atmosphere and rain forests…
- When to call an exploring child back within easy earshot…
- When to interrupt an enthusiastic employee whose project is moving ahead quickly but is starting to veer off-course…

These are difficult situations to judge. Excessive restraint can stifle enthusiasm and innovation. Excessive liberties can lead to chaos and crisis.

A key to effective regulation is deciding how closely to monitor activity, when to make changes, and what changes to make. These questions seem to be present at all levels of our lives, whether it involves parents in a household, teachers in the classroom, managers in a corporation, governments in our cities or nations, or the United Nations in a tense and troubled world. Life on Earth is infinitely complex with countless variables in its evolution making it impossible to come up with a rigid rule about what to decide in any given situation. Maybe the best we can do is to assess a situation and determine who is best suited to make decisions.

So the big question is this: When it comes to making changes in the complex nested structure of society, who should decide what?

Low-Level Decisions

Experts like Professor Jan Tinbergen (Dutch economist and Nobel Prize winner)[24] and American college Professor Howard Richards[25] agree that decisions should be made 1) at the lowest possible level, but 2) high enough to account for the concerns and well-being of people and groups affected by the decisions. For example, cigarette smoking is a personal choice, but in public places where one person's smoke affects other people's health, a person's right to smoke is best controlled by an authority that has the best interests at heart of everyone present. Likewise, a factory controls its own manufacturing process, but if its emissions waft across borders of several nations, creating acid rain and doing environmental damage, then those national governments together, or perhaps a larger regional council of governments are best qualified to set emission standards.

So this is the third principle of society's well-being:

Principle Three

Decisions are best made at the lowest possible level, but high enough to account for the needs and well-being of everyone who will be affected by the decisions.

Within society, decision-makers at the lowest level are the individuals (and nowadays, intelligent products such as computers as well). So most decisions in society have to be made at the personal level. That puts most of the weight where it belongs—on human shoulders well engineered to bear it.

In a biosystem like the human body, decision-makers at the lowest level are the cells and various molecules such as DNA, RNA, and hormones.

At the highest level of the human body is the brain, and it's the brain that has to make the decisions for the good of the entire body—ranging in scope from what to eat for lunch today, to what career path to follow for the rest of our life.

At the highest level of human society is the United Nations. Although it currently lacks the authority to make hard decisions in the best interest of the world, that will have to change rather soon. Nations will have to relinquish a share of their autonomy on issues that hold the future well-being of our planet in the balance. There's simply too much at risk to leave our global fate in the hands of contentious noble-savages and their self-interested nations representing narrow interests. In fact, it's rather insane to do so.

Legitimate Representation in World Society

According to UN official John Fobes, there's one quality shared by every good regulating body (corporate management, city and national government, and so on): The regulators of a group reflect the diversity of the group in terms of the people's culture, ethnicity, politics, religious beliefs, race, gender, and so on.[26] And that's why I believe the UN is the ideal body to become a world government today; it already has representatives from all nationalities, cultures, and world religions.

Marc Nerfin, a Swiss journalist, teacher, and former editor of *IFDA Dossier*, explained to me how the UN is given the impossible task to steer the world with its hands tied.[27] It's an instrument of governments. It only has the power they give it, which is little. So it can only work well when governments agree on things, which is rarely. The solution is not to dismiss or dismantle the UN but to fix it and give it the power it needs to help govern the world in a meaningful way. Mr Nerfin would like to see the UN evolve into three Chambers—the Prince Chamber (representing countries), the Merchant Chamber (representing multinational corporations, banks, and other economic interests), and the Citizen Chamber (representing people and their associations), which would hold the Prince and Merchant accountable.

Many people would like to see the UN become a world government, but most people familiar with the situation admit that a simple one-nation, one-vote system wouldn't be fair. It would give a small, poor country of illiterate citizens the same authority as a modern industrialized country. Canadian peace researcher Hanna Newcomb is among those who have suggested that each nation be given a weighted vote based on such conditions as gross national product, population, health and education index, assessed UN dues, and energy consumption.[28] In other words, the bigger role a nation plays in today's world, the more weight its vote would carry on world issues through an empowered UN.

However it's done, the UN now has to be given the authority to make decisions at the highest level—decisions affecting the macro systems of this planet. We can't wait any longer.

Gerald Mische, late president of Global Education Associates' New York office, told me that certain problems can only be solved at the world level: the arms race, the international debt crisis, terrorism, international drug trafficking, and environmental problems such as acid rain, pollution of the oceans, destruction of the ozone layer, global warming, and radiation fallout, to name a few.[29]

Keith Suter of Australia added that the common heritage principle—the idea that some parts of the world should be internationalized, or placed beyond the limits of national jurisdictions—is now well established in international politics, but it is applied only to such uninhabited areas as the seabed, the moon, and space. He says it's time to enlarge the Global Commons—the common heritage—to include rainforests, topsoil, the atmosphere, and other parts of the planet that are key to our survival.[30]

In short, we have to allow the UN to step forward very soon and bring order to the world.

Equality and Freedom

Americans extol the virtues of human rights, but healthy societies in the future will sustain a balance between the rights of individuals to be free and the rights of societies to be stable, by ensuring not just personal liberties, but equity and justice among all people.

Canadian peace researcher Hanna Newcomb uses a political model that determines how highly a nation rates equality on one hand, freedom on the other.[31]

- Autocratic socialists (communists) rate equality high, freedom low. Cuba, China, and the former Soviet Union are examples.
- Democratic capitalists rate equality low, freedom high. The USA is an example.
- Autocratic capitalists (fascists) rate freedom and equality both low. Examples include the former Nazi Germany and several African nations today with failing economies.
- Democratic socialists rate freedom and equality both high. Japan and most European countries today are examples.

That practical model suggests that the best-balanced governments are democratic-socialist in nature, such as those throughout most of modern Europe. The governments most vulnerable to problems are fascist, like Japan and Nazi Germany in the 1940s, in which industry and government form a tight alliance, forge a nationalistic agenda, and force the people to align to it or be ostracized.

Who Should Regulate the Internet

As the Internet evolves, political, social and economic issues are surfacing. For example, the free flow of information enjoyed by some modern countries (such as the USA) presents problems for societies (say, in the communist bloc or in the Islamic world) that wish to regulate information more closely to promote social stability. It would be dangerous at this juncture to assume that one approach to regulating information is better than another. Perhaps time will provide a global solution.

Meanwhile, the world network could develop in such a way that each individual, each organization and each nation or religion could tailor the information uniquely, within the framework set by the higher levels. And in my book, of course, the highest level should be an empowered United Nations. The UN, whose agencies already manage many of the standards in worldwide electronics and telecommunications, should be given more authority to operate the Internet, mostly to ensure its stability and international compatibility. Within that umbrella, each nation or religion or multinational corporation could establish its own standards, based on its own political, religious, or business model. As these macro systems overlap and cross-cut each other they will have to open dialogs to negotiate common ground and acceptable standards for all.

Within the macro systems, individual businesses and communities and other groups would further tailor the standards for their own purpose, always within the framework provided by the higher levels. Finally, most of the tailoring would be done at the personal level to fit the needs and desires of the end user.

A couple of experts at British Telecom told me that that is similar to the approach taken today by most of the major telecommunications companies.[32] These companies provide small networks (several hundred telephones and computer terminals hooked up to central computers) to companies, governments, hotels, schools, and other groups. Each of these networks can be tailored to the unique needs of the organization. Within that, each telephone and computer terminal can be further tailored to the specific needs of the individual who will be using it.

Carrying that philosophy to the global level—and carrying it carefully, with much discussion among today's diverse groups of people while it is happening—could provide the answer to the question, "How and by whom will world information be regulated?" Perhaps it will be regulated by people and groups at all levels, from personal to global . . . each person or group tailoring its own part of the system within the framework of the higher levels.

For now, many of us have very mixed feelings about how to address that question. Our noble side says, "Yes, we want protection for ourselves, and especially for our children, from the predatory peddlers of pornography, drugs, gambling, and other threats to our weaknesses that are so prevalent on the Internet." Our savage side argues, "I don't want some politically driven or profit-motivated authority dictating Internet content and probably charging me lots of fees! It's up to us to protect ourselves and our families from the smut peddlers. Don't tamper with the Internet!"

So, again, we need to move very carefully when addressing that crucial question: "How and by whom will world information be regulated in the future?"

Communication Skills

If two people in the same household or two nations in the same world can't or don't exchange ideas, information and feelings regularly, they gradually grow apart, and eventually become in some ways incompatible. While the Internet seems destined to promote smooth worldwide communication, a number of important issues have yet to be resolved. For one thing, if two individuals can't communicate clearly face-to-face, the Internet is not going to help.

Professor P.A. Amrung of Thailand taught me that if two people don't speak the same language, or if they do speak the same language but lack the skills and attitudes involved in expressing themselves clearly and tactfully, technology can't bridge the gap.[33] People need to learn those skills and attitudes as a prerequisite to peace in their lives and in their world. Children all around the world should be taught at least two languages. They should also be taught techniques to promote good listening and communication skills such as politely reflecting back what has been said from time to time to avoid misunderstandings. Without that inner development among human beings, our communication technologies are benign, or worse—dangerous.

The United Nations was founded on the belief that our future is a united future, according to one of my spiritual mentors, Jan van der Linden.[34] "The positive thought and meditation of millions of people who hold the idea of the UN close to their hearts can counterbalance the doubts and skepticism of those who at present are imprisoned in predominantly materialistic, selfish thinking," Jan told me.

Another spiritual teacher, Pat Mische, told me that life on Earth is 1) a journey in-beyond to the soul where we are all one, cutting through tradition, belief, culture, history, symbols and other illusions that divide us artificially, 2) a journey outward into the contentious world of opposing religions and nationalistic views, guided by the light of forgiveness, toward the unity of all life, then 3) a journey forward toward paradise—a new Genesis.[35] "Plant fruit trees even if you'll die before they bear fruit," says Pat. "Use your talents, skills, and insights to plant seeds now toward a new Genesis. Our astronauts and cosmonauts go into space as technicians and many come back as mystics.... We are not over the Earth but are part of this single cell."

My friend* and mentor Robert Muller points the way to the future. As Assistant Secretary-General of the United Nations during its first four decades, Robert developed a unique and wide-ranging worldview.[36] He says, "I believe that dreams are the surest way to new realities. My ultimate dream is to see this Earth preserved and improved as the beautiful paradise in the universe with a humanity living in peace, well-being, and utmost happiness in it.... How exciting will be the day when we will be able to say: We did it, we made the Earth a paradise, a garden of Eden inhabited by a happy human family! And how exciting this task will be for the 21st Century!"

Amen, Robert!

The main ingredients of good regulation are wisdom and foresight on the part of decision-makers, and wise regulators can make good decisions for a social system only if they have reliable information on which to base their decisions. For nations that means reliable economic information, the subject of the next chapter.

* Robert Muller, sometimes called "the UN's optimist-in-residence," is one of those fellows whose natural warmth and enthusiasm compels many people who meet him and spend time with him, or who work with him on a project, to regard him as a friend. That's my experience, having spent time with Robert on several occasions and collaborating on a couple of books—*Solutions for a Troubled World* and *Healing the World...and Me.*

CHAPTER 19

E-conometrics

The Geek Shall Inherit the Earth.
☺

IN THE PAST FIFTY years, computer scientists and engineers have transformed much of world society from thousands of isolated, suspicious tribes and communities and nations into one richly diverse humankind whose members are quickly getting to know each other, thanks to the Internet and the millions of personal computers attached to it. Thanks also in part to cell phones, TVs, movies on DVD, and other high-tech communication devices and networks. But nothing rivals the Internet for the transformation of world society through information.

Today, as information spreads quickly around the world, millions of everyday citizens in many countries know far more about foreign cultures, exotic religions, and distant lands than they knew twenty years ago...and if they *don't* know but *want* to know, they can click into *wikipedia.com* to find out in seconds! What a wonderful, massive library of information wikipedia is, compiled as it is by millions of free thinkers from around the world sharing their best knowledge and beliefs with everyone! While some misinformation and misconceptions find their way into the system, by far most of what I've read on wikipedia seems factual. In any case, the Internet today opens vast possibilities for unifying the world through the free sharing of noble-side information.

I envision a time, hopefully in the near future, when something that I call "**E**-conometrics" (emphasis on the **E**) will solve most of the problems facing humankind. The term suggests measuring and tracking economic variables through a computer network like the Internet. I adopted the term with the emphasized **E** in 2008 to describe a principle I've been refining since the 1970s, which I'll explain in a moment. For now I'll just say there are indeed economic variables involved—what I consider to be the three most basic and important variables—and they will revolutionize economics.

But first, don't confuse **E**-conometrics with its established counterparts—economics, macroeconomics, and econometrics (no emphasis on the *e*). If we go on the Internet to google those three terms, we soon start reading about such things as production possibilities, opportunity cost, regression analysis, simultaneous equation methods, and instrumental variables, and most of us quickly give up trying to understand them. Conventional studies of economics are frustratingly complex! Better left to the experts, right?

Well, that's a problem. The experts seem to disagree as often as they agree on what their complex theories and equations prove!

Today's economic systems (as well as political systems and, yes, religious systems) have evolved into an unnatural tangle of abstract concepts that give economists (and regulators and religious leaders) the frustrating task of controlling and measuring and judging and deciding things that are too unwieldy to be controlled or measured or judged or decided with existing tools and knowledge.

Basic Principles of E-conometrics

So let's sort things out, starting with economics. **E**-conometrics as described in this book boils economics down to a single, simple, untarnishable ratio that's easy to understand, and which I'll introduce in a moment. The following table compares **E**-conometrics to more established economic studies.

Basic principles of economics—a comparison

E-conometrics	Economics, macroeconomics, and econometrics
Tracks three clearly defined economic variables of society: people, products, and resources	Employs statistics and math for numerical analysis of both abstract and concrete economic forces in society such as interest rates, capital, and labor.
A nation (or any other social system) consists all of its citizens and all of the products they use—period—the basic building blocks. Territorial claims, borders, boundaries, and concepts of citizenship and ownership are abstract notions and are the concern of legal systems, not E-conometrics, which deals only in basic, substantial things.	Although sometimes regarded as a group of people with common heritage and culture (as in the Jewish Nation or the Cherokee Nation), a nation is more widely thought of as a nation-state—people living under one government within territorial borders (as in the 192 members of the United Nations). The nation consists of the people and their possessions and territorial claims. By that definition, a nation would include everything within the national borders, minus visiting foreigners and foreign-owned property, plus citizens traveling abroad, plus outside products that are owned by the citizens and groups within the nation....
Natural resources are the "food" of nations (and other social systems). They are outside the social structure of people and products (not necessarily outside the territorial borders), and they are useful and available to the system.	In traditional economics, natural resources were land, labor, capital, and entrepreneurship. Today at least three of those four things are called "factors of production" rather than "natural resources," and natural resources are usually defined as raw materials in the environment.
The economic stability and well-being of a nation are determined by the relationship (or ratio) between 1) system needs (what's needed to sustain the people and products), and 2) the resources available to satisfy those needs.	The economic stability and well-being of a nation are determined by highly complex variables, and economists of different persuasions (Keynesian vs. supply-side, micro vs. macro, positive vs. normative, mainstream vs. heterodox...) often disagree on which variables are most important.

Vitality Ratio

Although the variables (needs and resources) can be a bit tricky, *E*-conometrics can be summed up in this very simple ratio: "Economic Vitality equals Resources over Needs," or "V equals R over N," or:

V = R:N

It simply means that the ratio between resources and needs is the main factor determining the economic vitality of a social system, just as the food that a biosystem eats is the main determining factor in the health of the biosystem. If there are enough appropriate resources to satisfy the needs of the social system (of people and products), then the system is healthy. If there are resource shortages, the well-being of the system begins to diminish. So this is the fourth principle for a society's well-being:

Principle Four

The ratio between needs and resources largely determines the overall health of society.

Simply plug that ratio (V = R:N) into the Internet, and away we go toward a healthy, prosperous future for all humankind!

Obstacles to Implementing E-conometrics

Well, it's not *quite* that simple. There are some obstacles in the way to getting **E**-conometrics working in today's world. (This section gets a bit detailed, and I'm sure some of you will want to gloss over it or skip it entirely, coming back to it later, if and when you're compelled. For you brave souls who don't mind the detail, please read on!)

First of all, many people are emotionally attached to certain economic or political or religious or scientific beliefs that might be incompatible with **E**-conometrics, so there will certainly be some resistance to getting it on its feet.

Also, there has to be a clear delineation between the social system (people and products) and the natural resources. Great care has to be taken in discerning products from resources. For example, a mushroom grown in a domestic greenhouse is a product. Growing wild it's a resource until it's processed, then it's a product. Grown in a foreign greenhouse it's a resource until it's imported, then it becomes a product. The status of every product and resource has to be tracked in detail. A natural analogy to this is how Vitamin D in the body can be either a hormone (if produced in the body) or a vitamin (if produced outside the body and ingested as a nutrient, as in a slice of cheese). The body's hormones are akin to a nation's products, and the body's vitamins are akin to a nation's resources. All quite natural...but somewhat complicated.

Another complication: Theoretically **E**-conometrics can apply to any social system of any size, and that can get tricky. The natural resources of a family would include the things they buy at the store, and once those things are brought home and put away in the fridge or pantry, they're products. Those same things on the store shelves that are resources of the family, are actually products of the city and the nation where the family lives.

This is a quirk of living on Earth because of life's nested structure (systems within systems within systems), as we explored earlier. Think of a biosystem eating something. In fact, let's say *you* are eating an apple. To you, that apple is a food (a resource) until you swallow it. Then it's broken down into tiny nutrients including molecules (products) that are distributed throughout the body and ingested by your body cells. To a body cell, the molecules it accepts from the system are food (resources) until they enter the cell to be used, at which time they become products of the cell. To the society, the apple is a product until you eat it; then it's a part of a person—you.

So that's one of the complications in implementing ***E***-conometrics—coming to grips with the nested structure of life. To keep things manageable in this book, from this point forward, nations are the social systems we'll be most concerned with, as they have become the predominant life form on the planet. So the focus will be nations rather than the industries, cities, neighborhoods, friendships, clubs, religious groups, and other social systems that make up the nations.

To implement ***E***-conometrics we would have to start with nations that are already integrated with a modern infrastructure—communication and transportation networks (especially a well-spread computer network); electricity, food and water readily available to everyone, and so on. We couldn't implement ***E***-conometrics in a primitive, nomadic society, for example.

Now, this is the most important obstacle: We can have only limited success trying to implement ***E***-conometrics in single nations. It can't be completely successful until it's implemented at the world level, bringing all nations into a single, integrated society of humankind. The reason is that people and products move fluidly among nations in the course of tourism, trade, and migration, making it nearly impossible to keep the "needs" variable steady. For example, one nation might implement ***E***-conometrics and keep its birthrate at a safe level, while other nations nearby let population grow out of control, compelling the crowds to overflow into the more stable nation, thus destabilizing it.

So ***E***-conometrics ultimately will have to be implemented at the planetary level, which will mean all nations will have to be fitted with modern communication, transportation, and energy infrastructures. That's the first step. Then, when all nations are up to speed, the global network will be implemented and monitored by many nations and corporations working together through the United Nations.

Planetary ***E***-conometrics can be accomplished only through an empowered United Nations, when the noble side of humanity is worked into the formula. Excessive needs will be eliminated through wise planning throughout the world, everyone's basic needs will be met, and only within that balance will individual nations be able to live out their economic, political, and religious compulsions.

Businesses in rich countries can enjoy a free marketplace only to the point where their freedoms begin to upset the balance of global needs and resources; then they'll have to back off.

Families in poor countries can enjoy having lots of kids only to the point where population growth in their country begins to outstrip resource availability; then they'll have to back off.

Our new world, integrated by **E**-conometrics, needs to flourish with nobility—cooperation, sharing, and the good will toward all. Such values come naturally only to our noble side, and that is the side of humanity that will have to sustain **E**-conometrics through the United Nations.

Only within that stable system of noble values, then, can macrosystems (nations, world religions, and transnational corporations) maintain their own political, business, and religious systems...and only to the extent that they don't upset the noble balance.

In short, economic, political and religious dramas stirred up by incompatible ideologies and dogmas cannot be allowed to sneak into the **E**-conometrics network. The emphasis always has to be our noble side, which rests on compatibility and the best interests of all humankind and the planet.

Will today's macrosystems give up their autonomy to a noble world network? If not, then that's undeniably the biggest obstacle we face.

Technical Considerations

Finally, the technical obstacle to implementing **E**-conometrics: The elaborate, high-speed computer network will have to keep track of the variables in exhaustive detail—but don't worry; it's not *you* who'll get exhausted; it's the computers and the computer experts who will do all the work and, quite literally, transform the world by their efforts. These are some of the variables they'll track:

A nation's system needs (population and demographics, per-capita consumption of products, recycled products, product life expectancy, products in use, products in storage in warehouses, products on store shelves, products stored in homes and offices, nutritional qualities of consumable products, wholesome vs. unhealthy consumables vs. medicinal consumables etc.). Eventually every home, office, and school will probably keep a running inventory of all products they acquire and use, but that would probably be too much to ask at first, so the lowest level of reporting initially could be the retail merchants who sell products to families, offices, schools, and other end users. Most of them already keep sales figures and running inventories that could be plugged into the system.

Resources (reserves of raw materials owned by the nation and its people, foreign raw materials accessible to the nation, renewable vs. nonrenewable resources, imports of foreign products, natural energy such as sunlight and wind, and more).

Never before in history has it been possible to track so many widespread variables accurately. Now, thanks to computers and network technologies, it is. Because of that, ***E***-conometrics can move humankind ahead more quickly toward paradise on Earth than ever before possible. We could do it this century!

Now how do you feel about this so far? Are you excited by the prospect of unprecedented social stability and being on the threshold of paradise? Great! Me too!

Or are you at least heartened by the possibilities of satisfying the basic needs of all people on Earth as a first step? I am.

Or are you a bit intimidated by the massive scope of the project? Yes, well, I am too. But I've worked in high-tech companies since the 1970s, and I know that a project of this type is well within the technical range of possibilities in today's world.

Or is a little voice within you setting off an alarm about invasion of privacy or a threat to your beliefs or freedoms or…? If you hear that voice of alarm, it's okay. It's your savage side reacting with mistrust, and it's a natural part of us humans living on Earth. A safety mechanism. Just bear in mind that caution and uncertainty can make life on Earth an adventure, but poisoning doubts and fears can spoil the best works of humankind and paralyze one's life. I'm once again getting off onto a tangent, though, so let's get back to **E**-conometrics and how to implement it.

Internal and External Affairs

Other factors also have to be monitored, including the impact the nation exerts on the living systems around it, especially ecosystems. The terms "domestic affairs" and "foreign affairs" should probably be replaced by "internal affairs" and "external affairs." A nation's internal affairs are everything related to people and products. External affairs include not just natural resources, but also the ecosystems, foreign nations, other living systems external to the nation's structure of people and products, and of course, the United Nations organization.

Remember, we're not talking about political borders at all; society is the people and products, period. So, in a town, for example, the houses and yards are products. Gardens and parks and golf courses are products. Streets and stores are products. But many things in the city limits are *not* products. Empty lots overgrown with weeds are ecosystems. Although they're inside the city limits, they're outside the system of people and products that compose the town. The surrounding ecosystem has many inhabitants who visit the social system as welcome guests...or sneak in to cause problems—squirrels, birds and butterflies on one hand, rats, mice, houseflies, and spiders on the other.

Enough said. That gives an idea of how social systems (towns and homes in particular) are not defined by borders, but by the living structure of people and products. Take a moment to think about how other social systems (states and provinces and nations...) are structures of people and products, and how ecosystems and biosystems can interpenetrate those systems even though they are not part of the social systems. Think of people sneaking into ecosystems to kill deer and elk, and think of cougars and foxes sneaking into towns to kill pets in back yards. Think about how massive world religions like Christianity, Islam, and Buddhism are also social systems composed of people and products (churches, religious texts, icons...), and how they overlap and cross-cut with nations. Think about how some nations (the USA, for example) strive for a separation of church and state, while others (especially Islamic countries) revel in the entwinement of religion in all aspects of society, including government.

Think about all of these social systems, not as political borders or ideologies or races or dogmas...but as living systems composed of people and products. Before long it gets easy to see this new, more natural view of society and humankind, and life on Earth begins to fall into perspective more clearly in our minds. This, in my view, is the first step toward a paradise existence on Earth. We begin to see clearly that we are of nature, not above it.

Money and Life Energy

No discussion of economics would be complete without the mention of money, so let's put it in perspective. Money flows through our world, for better or worse, as a symbolic substance representing people's skills and services, products, natural resources, and other things that people consider important. When put to use by our noble side, money can be used on Earth to materialize our boldest, most beautiful dreams. Fueled by our savage side it can bring on our worst nightmares. So money is not "the root of all evil," nor is it a panacea for all problems. It's simply a force, a potential, to help us manifest our dreams and intentions, for better or worse.

Money is the physical world's version of life energy or *prana* or holy spirit, which flows freely through the worlds of spirit, nourishing all life. Life energy materializes the desires and intentions of all beings instantly, making dreams come true and manifesting miracles. That's the basic way of life everywhere....

Everywhere but the physical realm. Subtle life energies nourish the living spirit within us, but our physical body seems to be impervious to their nourishing, healing effects and their powers of instant manifestation, unless those imperceptible forces are transduced through the receptive spirit inside us. And that transduction process (pulling life energies from our spirit bodies, through our physical body, and releasing them into the physical world) can usually occur only when we go through a spiritual development process such as practiced meditation, in which we foster a rich connection between the conscious mind and the finer spiritual mind.

Otherwise, life energy does not manifest reality in the dense physical world as it does in the subtler worlds. Money does. But enough said about money.

Maintaining a Balanced Ratio

The aim of *E*-conometrics is simple: To sustain a balance between needs and resources. In the course of human evolution the usual trend has been rising needs. As population increased and societies became more complex with greater varieties of products, the needs for resources continually grew. So, maintaining a balanced ratio in the future will involve finding ways to reduce needs and to increase resources in safe, healthy ways. Two of the most crucial factors will be renewable resources and population control.

Renewable resources. Hilkka Pietila, head of the UN Association in Finland, advocates moving away from an extraction or "dead" economy (taking nonrenewable, "dead" resources from the Earth) toward a cultivation or living economy (interaction with living, renewable resources).[37] Conventional economics doesn't make enough of a distinction between renewable and non-renewable resources. Oil, coal, and natural gas are nonrenewable. They simmered and fermented underground over millions of years—it took the decay of 98 tons of ancient forest to produce a gallon of gas today—and once those resources are used up they're gone forever. Today they're quickly disappearing, and competition to get them is growing fierce. On the other hand, energy from the sun, wind, tides, and geothermal sources are renewable; they're ever present on Earth. We need to rely mostly on renewable resources, says Ms Pietila, but modern economics doesn't address that need.

E-conometrics would take into account all known factors pertaining to resources and needs, always pointing us toward wise choices, including an emphasis on renewable resources.

Controlling population growth. Overpopulation has been a perennial problem in the Far East, and some societies there have learned to deal with it. The Japanese in the 1950s reduced population growth to about 1 percent a year and emerged in the later decades of the 20th Century as a leader in world industry and technology. Other once-struggling East Asian nations enjoyed similar economic rewards. Singapore and Hong Kong cut their population growth rates in half during the 60s, and followed on the heels of Japan in world trade. If poor countries were to enact effective population policies now, in the opening years of the 21st century, presumably they would begin feeling economic vitality by 2050.

Let's take a look at the master of population control—China—to get an idea of how it could come about. The sprawling nation of China comprises a sixth of the world's population. China entered the Cultural Revolution in 1949, driven by the exuberance of an ancient culture transformed to youthful vigor by new socialist ideals—the public ownership of property and the reverence of social equity and stability over personal freedoms.

State Family Planning officials Dr Liang Jimin and Wang Xiangying told me that, economically speaking, the Cultural Revolution was a period of two steps forward, one step back.[38] Population doubled from a half-billion to a billion from 1949 to 1987. By 1982 80% of the people were semi-literate (less than middle-school education) and lived on farms, and it became clear to the leaders that calamity would strike soon if something weren't done. A comprehensive network of family planning was established, touching every Chinese household. The family planning network made it possible to instill in everyone the importance of later marriages (mid-20s), healthier pregnancies, and just one or two children per couple. Quite a switch for a culture in which the average family had 5-6 kids in the 60s, 4 in the 70s, and 2-3 in the 80s.

The Chinese stuck closely to their population plan, and today the people enjoy the most prosperous lifestyle of their long history. Had they not followed the plan they almost certainly would be contending today with poverty, famine, and other symptoms of a negative ratio, as they have done so many times throughout history. The Chinese still have their share of economic problems, but they're small in comparison to the famines that ravaged their society in bygone eras.

Humanity as a whole is not so fortunate today. Population in much of the world is out of control. As we quickly approach 7 billion people on Earth I believe we will soon see devastating symptoms of unprecedented proportions in many parts of the world unless we can get a handle on population growth very soon.

Causes of a Low Ratio

Anything that causes the needs of a social system to increase (growing population or rising per-capita consumption, for example) and anything that causes the resource availability to decline (natural disasters, depletion of non-renewable resources, or resources lost by war, for example) can result in a low ratio in which needs exceed resource availability. Here we look at two of the leading causes today—uncontrolled population growth and growth economics.

Overpopulation. Of all the variables involved in needs and resources, none is as crucial as human population. Overpopulation has probably been the most pervasive negative ratio condition of humankind down through the ages, simply because we humans are programmed to love sex and to have lots of babies. It traces back to our ancient cross-breeding for survival on Earth.

Experience around the world has revealed many devastating symptoms of overpopulation, including famine, war, environmental destruction, and mass execution. Again, there will soon be 7 billion people on Earth, and devastating symptoms of unprecedented proportions are likely in many parts of the world unless we can get a handle on population growth very soon.

Recent tragedies in the heart of Africa could be a taste of things to come in much of the world. Before we explore those tragedies, let's look at some of the hotspots and not-so-hot spots in the world today.

Country	Population, 2006 (from www.photius.com)	Births per 1,000 population, 2008 (from www.wikipedia.com)
Germany	82,400,996	8.2 (safe)
Singapore	4,553,009	8.2 (safe)
Japan	127,433,494	8.3 (safe)
China	1,321,851,888	13.1 (safe)
USA	301,139,947	14 (safe)
India	1,129,866,154	23 (hot)
Rwanda	9,907,509	44.5 (very hot)
Uganda	30,262,610	46.6 (very hot)
Congo	65,751,512	49.6 (very hot)

This table gives a general idea of where the population time bomb ticks and is liable to explode in the near future. The main factor to watch is birthrate. Large numbers (40+) suggest large families—a condition in which population will try to double in less than a generation. I say *try* because there are always limits to growth, and those limits trace back to resource availability. When the limits are reached, symptoms begin—famine, mass execution, mass migration, and so on—bringing population back into line, often in brutal ways. But I'm getting ahead of myself.

India is one likely hotspot, with an immense population (1.1 billion) and a high birthrate (23 births per year per 1,000 people). There are conditions in India that alleviate the problem there, however. The people are well-educated, so many of them move out of country to find jobs elsewhere in the world. Of course, while that relieves the economic pressure in India, it compounds the problem in the nations who receive the large streams of immigrants from India.

A more serious hotspot is in the heart of Africa, where Uganda, Rwanda and The Congo come together. A history of large families has led to perpetual overpopulation interrupted only by a long series of tragedies—famine, civil war, mass emigration, mass execution, and environmental destruction.[39]

Straddling the three countries is a national park—actually three adjacent parks with some of the most incredible biological diversity in the world: Virunga Park in Congo, Gorilla Park in Rwanda, and Mgahina Park in Uganda. Some 400 mountain gorillas inhabit the parks, more than half of the world population of mountain gorillas. Until recently tourists have flocked to the parks to see the dwindling populations of mountain gorillas, bringing in $80 million a year to the economies. Tourism has fallen off recently because of serious problems stemming from chronic overpopulation in the region. Symptoms incude:

- The environment is being pillaged. Nearly everyone living nearby uses charcoal to cook their food, so with more than 1 million people living along the fringes of the park there's an immense, growing need for charcoal, which comes from clay kilns built in the forest wherever trees are felled...which is occurring over three vast areas within the park. Long queues of women carry on their heads 150-pound bags of charcoal over winding forest trails into the cities, where they are distributed on vehicles and sold throughout the region. Rangers try to prevent the illegal charcoal harvesting in the park, but society's burgeoning demand generates incessant pressure to get more at any cost. Militias try to protect the people's livelihood by threatening and if necessary killing the rangers who protect the park. So acre after acre of forest in the park is decimated. But harassing park rangers is the least of the militias' savage agenda.
- Perpetual tensions between the Hutu and Tutsi people came to a head in Rwanda in 1994 with the mass execution of 800,000 Tutsis, perpetrated mostly by Hutu militias.
- Civil wars in Congo between Hutus and Tutsis from 1996 to 2003 killed 5 million people, the largest war casualties since World War II. Those inter-tribal wars continue today, stirring soldiers' savage side, compelling them to rape, torture, murder, and pillage as they move through village after village.

How to resolve the gruesome problems in Africa and elsewhere? Some people would send money to ward off famine, but that doesn't work. Money is absconded by the cancerous kleptocracy—government by stealing—that has spread through much of Africa. And support that *does* get through to enrich the pathetic lives of the starving masses, alleviates immediate symptoms while enabling families to continue to grow. So adding money and food to the situation escalates the problem in the long run.

Some people would send peace-keeping troops to keep order, but it's like trying to keep order among hungry scavengers around a fresh carcass. As long as there's desperation there will be conflict and suffering, and as long as there is overpopulation there will be desperation.

There's only one noble solution—a family-planning program like China's. Can it work in Africa? I'm certain it can under the right conditions, which would include integrating the region with networks for renewable-fuel-based transportation, electricity, and communication—especially a computer network—and implementing ***E***-conometrics.

And I'm confident that we could resolve these grave symptoms in the span of 50 years in the region surrounding the African parks, a hotspot that might be the perfect beta-site, or test site, for ***E***-conometrics. For one thing, the region is in dire straits with little to lose and everything to gain. Also, there's a lot of volcanic activity in the area, suggesting that geothermal resources could be harnessed for a limitless supply of cheap energy. Engineers from Iceland could be recruited to develop the electrical generating systems. Not only would they have the technical know-how (geothermal energy production is well-developed in Iceland), but they'd certainly have the incentive to get air-conditioning systems up and running in a hurry in that sweltering equatorial environment! ☺

A modern infrastructure of transportation and communication could be developed with the help of teams from North America and Europe.

Chinese experts could be recruited to help develop a nested family planning and education organization throughout the area to raise the literacy rate. I realize there are racial differences, and I suspect it might be easier for Orientals to adjust to a nationwide family planning program than for, say, blacks, whites or Hispanic-Latinos.* Still, I'm sure it could work for all healthy human beings today under the right conditions.

Finally, Africa's the continent that's been abused and neglected down through the ages—the source of slaves and the target of Euro-colonialism. This would be a chance for Africa to move quickly to the forefront of human society, showing the way of the future. Many cultures have had their moment in the Imperial sun: Akkadians, Egyptians, Indians, Babylonians, Chinese, Hittites, Assyrians, Greeks, Romans, Mayans, Incas, Muslims, Europeans, North Americans....

Maybe now it's Africa's turn. Another reason this would be an ideal beta-site for **E**-conometrics: the three parks make this area

* My mind keeps going back to the opening ceremonies of the 2008 Beijing Olympics, when large blocks rose and fell like ocean waves with computer-like precision, and scores of Chinese men then climbed out from beneath the blocks to reveal that the flawless execution was the result of complex human choreography, not computer programming. I had a difficult time envisioning white, black, or other non-Oriental men emerging instead from that meticulously synchronized performance.

the jewel of the continent, maybe of the world. It could be the showcase for the new human paradise, there in the heart of Africa!

Implementing *E*-conometrics and coupling it with an intricately nested family planning program like China's would remove most of these symptoms from our world once and for all, in the course of several decades, but it will have to be implemented rather soon to avert impending disaster. Let's give it a shot! We could begin in Africa.

Growth economics. While overpopulation is the main economic problem in poor countries, in some wealthy countries the culprit is high per-capita consumption and the growth economics that pushes it along. Modern economic thought is based on the belief that economic growth is the main measure of economic health and vitality, but it is a false and dangerous belief. Unbridled growth (which in a biosystem is called cancer) is simply a boiling-over of our savage compulsions to acquire more-more-more...more land, more food, more luxuries, more money. Biosystems like the human body grow physically until they mature, then they sustain. That's what healthy societies would do. The always-grow-and-never-mature economic priniciple might have been important in the past in our drive to spread order out into the chaotic ecosystem by converting more and more land from ecosystem to social system, but today, as swelling nations push up against each other in the global ecosystem, the economics of growth breeds mistrust, conflict and inequity throughout most of the world.

Before my spiritual quest (when I was trying to heal the world before I'd healed myself!) I spent several years collaborating with bright minds from various countries to try to identify and solve the biggest problems plaguing our world. Several of them talked about unbridled growth.

- Ahmad Abubakar of Tanzania referred to Jan Tinbergen's "Report to the Club of Rome," which pointed out that industrialized countries consume 20 times more per-capita resources than do poor countries.[40] Mr Abubakar warned that we humans—as individuals, as nations, and as a species on Earth—need to reexamine consumption levels to eliminate waste, to divert wealth away from military research and redirect it toward tackling global problems such as disease, hunger, and natural disasters. That change of direction can't be achieved by a group of distrustful, inequable nations, says Ahmad Abubakar; it can only be done by turning the UN into a world government so that nationalism takes a back seat to globalism.
- The notion that ever-increasing wants should be continually satisfied by ever-increasing production is leading to the suicide of civilization, warns J.S. Mathur of India.[41]
- Howard Richards warns that our purchases of more, more, more provide a short-term "fix"—a momentary jolt that makes things seem better—but offer little long-term satisfaction. They just offer short-term relief from the cravings. In other words, money and growth can become an addiction.[42]

So nations in the future will have to focus on economic sustainability rather than economic growth. The only way I know to ensure sustainability is to maintain a balanced **E**-conometrics ratio.

Symptoms of a Low Ratio

When the *E*-conometrics ratio goes negative—when needs exceed resources—all sorts of economic problems can develop, some simple and short-lived, others devastating and long-term. Symptoms of a negative ratio include:

Fewer products per capita. Needs for particular resources exceed supplies. As a result, there are fewer products made from those resources—fewer products to go around.

Rising prices. Carnivores during a drought fight more aggressively over a carcass, trees in a dense forest grow as tall as possible to compete for sunlight, and social systems facing a shortage of a particular resource pay more money to get it and its related products. Freezing or flooding or drought can ruin thousands of acres of raw farmland in any given year, resulting in shortages of wheat or rice or soybeans or oranges. Like the toughest carnivores and the tallest trees, the highest-paying social systems (stores, processors, etc.) get the goods. When resources (in this case, fertile farmland) are insufficient to satisfy needs, expensive products spread through society, and prices rise.

Inflation. Often called "too much money chasing too few goods," this symptom could more aptly be called "too many needs chasing too few resources." As people and groups pay higher prices for the scarce resources and related goods, they demand more compensation for their own goods and services, and prices spiral upward.

Recession. As inflation spirals and things grow scarcer and get more and more expensive, it gets harder for social systems like companies to keep doing what they do, so things start to slow down. They cut jobs and maybe close their doors. This is recession, which often follows on the heels of unchecked inflation. And again, recession can usually be traced back in time through the inflation, to a negative ratio in which needs exceed resources. Recession is an unwitting effort by social systems to reduce their needs.

Depression. If recession doesn't adequately reduce needs, depression follows. As the unemployment lines grow and more commercial-industrial organs die within a nation, the surviving social subsystems and the nation as a whole begin to weaken dramatically, like an old man on his deathbed. As more businesses fold and the nation's physical structure continues to decay, products are being manufactured and distributed in inadequate numbers. Resources may be growing plentiful, but the nation has no way to digest them, so they are not really resources anymore...just as food is no longer really food to a dying man. The nation is on the verge of depression. It is dying. Fortunately, nations are not yet mature living systems. When nations "die" during a severe depression, they can rebuild, hopefully having learned from their mistakes.

The preceding symptoms of a low ratio are usually experienced by more advanced nations with a growth economy and can usually be traced back to needs outstripping resource availability. They could be eliminated by ***E***-conometrics, which would raise a red flag as soon as needs begin to exceed resources, and a series of options (cutting back on particular products for awhile, finding replacement products or resources, or acquiring more resources from specific sources, for example) would be offered to help restore the balance.

Those symptoms are most debilitating to advanced nations whose physical structures of people and products have grown fairly complex. Poor nations are not as vulnerable to sophisticated symptoms. Their needs are different. The usual cause of a low ratio in poor countries is overpopulation, as mentioned earlier, and these are among the most common symptoms:

Famine. Primitive cultures and other poorly integrated societies don't have a diversity of products. They need a steady supply of resources to feed the people, but only a modest amount to sustain the humble infrastructure. So, the usual cause of a severe resource shortage in a poor nation is overpopulation, and the chief symptom is famine. While the elaborate infrastructure of the advanced nation crumbles, poor nations are riddled by starvation and disease when their needs outstrip resources through overpopulation.

Mass execution. When resources are in serious short supply, envy and desperation often lead to gross inhumanity. Mass execution is an unconscious, desperate effort by factions in a nation to solve economic problems by reducing needs. Just as a man whose family is starving might steal or even kill to feed them, a nation suffering a severe imbalance between resources and needs often vents its frustrations in cruel and unjust ways. The targeted victims of mass execution might constitute a group within society that is unwilling or unable to conform to national objectives or regulations for such reasons as religious belief, ignorance, intertribal contentions, or geographic isolation. Through mass execution some nations attempt to solve two problems—reduce needs and dissect an incompatible segment from the national structure.

Mass emigration. Occasionally there is an outpouring of people and products from a particular nation. Whether the group is exiled or feels pressured to flee for political or economic or religious reasons, it usually happens when the nation is suffering economic hardships—or, more specifically, when resources are in short supply. In the last half of the 20th Century, Africa had 5 million homeless, 125,000 Cubans fled to America in a "freedom flotilla," 800,000 Afghans fled to Pakistan, 500,000 Vietnamese fled to Thailand, tens of thousands of Jews fled from the Soviet Union, and hundreds of thousands of Mexicans poured into the United States. When mass emigration occurs, needs are reduced in the nations left behind, and the receiving nations take on the economic strains of rising needs.

Those three economic syndromes of poor countries could also be eliminated by **E**-conometrics, whose aim, again, is to sustain a balance between needs and resources. In a country prone to overpopulation, needs would be kept in check largely by a multi-level family planning program like the one that transformed China from a peasant economy to an industrial leader in the closing decades of the Twentieth Century. A family-planning program, along with education (and, of course, an infrastructure of transportation, communication, and electricity), would be the backbone of **E**-conometrics in poor countries.

The last two symptoms mentioned here, below, can afflict any nation, rich or poor, when needs outstrip its resources.

War. Like mass execution, war is often a desperate attempt by a nation to bring needs into line with resources. It's often waged to steal resources from another country, such as oil in today's world. War also reduces needs by removing many people from the equation—military and civilian casualties.

Ecological destruction. When needs exceed resources, nations often become desperate enough to exploit the environment ruthlessly for more resources. When a nation becomes desperate, environmental concerns often take second seat to keeping the bloated structure well-fed, especially when leadership is weak or misguided. Land is ravaged, water and air are poisoned, and life cycles in the ecosystem are upset or devastated.

E-conometrics would eliminate war and environmental destruction along with the other symptoms by making sure needs did not exceed resources. I know it will work.

The Ethereal beings gave us world-changing information between 1995 and 2000, then backed out of the picture to see what we humans would do with the information given so far. I'm sure they will come close again to offer support and guidance to humankind in much more profound ways, once we get our house in order. And I believe *E*-conometrics will help us get our house in order as nothing else can. After all, that's what *E*-conometrics means (from its Greek roots): oikos (house), nomos (custom or law), and metron (measure) or "measurable rules of the household." And the emphasized "*E*" suggests letting computers do most of the work! Earth is the ultimate human household, and *E*-conometrics is the only way I know to get it in order fast.

The Ethereal beings told us they've come close to our world six times in the past, trying to guide humanity toward a bright future. Each time they chose a civilization or society of people to work with, and each time the humans failed to rise to the occasion, and humankind fell into a dark age. Now the Ethereals are here for the seventh time, and I've finally determined who the chosen people are this time. For the first time ever on this planet the Ethereal beings have the opportunity to work with the society they've always wanted to work with, and that is ALL HUMANKIND. The Internet makes that possible for the first time ever!

We all know how the Internet has transformed the world in the past twenty years. **E**-conometrics is simply the next step. It'll carry us well along the path to Paradise—Heaven on Earth. Although it will work to full potential only when implemented worldwide, it could begin on a trial basis in areas where immigration could be controlled, and where disruptive groups opposing the principles of **E**-conometrics could be kept at bay, during the trial period of, say, twenty years. After a successful trial run, it could be implemented by all interested nations...and then ultimately at the global level, by a world government or by the UN...which hopefully will soon become a world government.

Trouble is, we're already well on our way toward the next End Time, with the meltdown of the polar icecaps, widespread flooding, devastating storms and natural disasters, destruction of the rain forests, international contention for the last drops of oil...but most of us remain oblivious. We humans tend to get caught up in the dramas of our day-to-day lives, and by the time we really notice the impact of the approaching End Times on our personal lives, it may be too late to turn back.

This is the time when we must all open our minds and hearts to our beautiful world in crisis. Let's give **E**-conometrics a chance.

CHAPTER 20

The Next Frontier

I WAS IN A SPIRITUAL research group that ended for religious reasons. Some members thought they were God, others disagreed. ☺

Now, that can be taken two ways. From the ego's point of view it's a joke suggesting that certain antagonistic individuals had an inflated opinion of themselves. From the spirit's point of view we are indeed God at the soul level, where we're all one...so those who *disagreed* were the antagonists. So, take the joke however you like; there's truth in both interpretations, depending on how you look at it!

Most of life flourishes beyond the perception of our five senses. Scientific instruments can delve far outward and deeply inward into unseen worlds, revealing the nature of distant galaxies and subatomic particles, but even science has its limits. With our senses and sciences together, what we perceive is still only a tiny subset of reality. Most of life flourishes in-beyond of us.

To perceive in-beyond we have to foster what is sometimes called our sixth sense, or our extrasensory perception (esp), through such practices as meditation and mind control. These techniques help us close down the five senses while remaining conscious, so that our higher senses open up. The more we do it, the more we are privy to wonders beyond our world, beyond our imagining.

The Spiritualization of Humankind Today

In the coming years our world will undergo a spiritual renaissance. Humanity at large will embrace the basic spiritual truth that rests at the core of every time-proven religion and esoteric school. That truth is what I include as Principle #5 of humanity's well-being:

Principle Five

Yes, there is a source of all life, and at the center of our being we are each a part of that all-knowing, omnipresent source.

As humanity embraces that principle, various myths and dogmas—many of them incompatible with each other—will continue to sprout from the core, live a rich life of several centuries or millennia, and later blow away like autumn leaves, but the core principle will stand the test of time as it always has, for it is timeless. Someday soon everyone will acknowledge its truth.

There's a big obstacle in the way of this global rebirth, however. Many large, powerful social systems are committed to agnosticism. Among the largest and most powerful is science. This may sound a bit dramatic, but unless we can persuade scientists to embrace the larger reality of spirit in the near future, it may be impossible to avoid the coming End Time. Finer spiritual sources have given us serious warnings about certain trends on Earth, but science blusters ahead, unheeding, with super-colliders and other potentially devastating technologies.

One day in the future, a scientist will develop some new piece of equipment to delve into the realm of waves and particles. She'll probably be sitting in a lab somewhere, and as she gazes into the equipment for the first time she'll see vast crowds of bright, smiling faces beaming at her, including her late parents. They'll be greeting her by name with happy voices that boom through the equipment. She'll probably fall off her stool.

Such a scenario might be the start of true, lasting ITC—a marriage of science and spiritual research. It will happen once science has the knowledge and determination to explore in-beyond. For now, though, scientists (not by choice, I think) make an effort to stay "out of tune" with the worlds of spirit, similar to the way a radio stays out of tune with stray radio signals.

When we turn on a radio we're not inundated by thousands of stations that are all sending their signals into the room at that moment, because the radio has a tuning circuit that allows it to tune into one station at a time. Occasionally we get bleed-through from two or three stations, it gets messy, and so we adjust the frequency to tune into a single, good station.

Our mind is like a radio. It has access to many of the nonphysical realms around us, but our five senses and reasoning faculties act as a tuning circuit, keeping the mind tuned in to a single world—this physical world. When we get bleed-through from the other realms, our five senses sharpen, we reason away the phenomenon, and our mind is then locked in to the physical world. For example:

- When we hear voices in our head from these other worlds, our ears perk up and we ask ourselves, "Where did those voices come from? Outside? The next room?" A psychiatrist might ask additional questions. "Was it a misfiring of my neurons? A biochemical imbalance?..." By perking up our ears and focusing on reasonable explanations, our mind becomes locked in to the physical realm, and under normal conditions the other-worldly voices disappear. (Under abnormal conditions such as schizophrenia and the use of mind-altering drugs, the voices might persist.)
- When we see movement out of the corner of our eye as a result of nonphysical beings passing through our world, we turn our head quickly, peel our eyes, and wonder, "What was that shadowy motion? Hm, just my mind playing tricks, I guess...." As we focus our eyes and rationalize the movement, we hone in on the physical realm, and the shadows disappear.
- When we awaken from a vivid dream in which our mind has been out exploring other realities, all details of the dream are swept away into the unconscious as the five senses kick in and lock us into physical reality.

This is how our mind works. It wants to stick to the business at hand—the goings-on of the physical world around us—so it blocks the spirit worlds from our perception.

For the past 300 years, since the days of Isaac Newton, science has excelled at exploring this physical universe and remaining locked into it. Laboratory conditions, the scientific process, peer review, replication of results, rigorous testing, and systematic investigation are among the techniques employed by science today to make certain that everything it tests remains within or slightly beyond the range of its understanding—within the parameters of the physical universe. That has allowed science to excel at exploring the physical domain, but it has caused a perceptual block of the nonphysical domains, so that any evidence of them, no matter how compelling, is rejected.

Now, back to that scientist who fell off her stool. Much will depend on what she does after she gets up and regains her composure. If she adheres to the scientific process—trying to replicate the results through rigorous testing, soliciting the opinions of peers, and so on—chances are she'll talk herself out of the experience, attributing it to delusion or stray TV signals or some other rational explanation.

But what if she tries something new? What if she cries for joy and allows herself to feel the presence of her parents there in the lab and the warm feelings in her heart? What if she holds onto those feelings throughout the day, and that evening, before bed, she states her love for her mom and dad in a prayer? What if she then, later that night, has a most wonderful dream in which she meets with her parents in the home she grew up in? Mom and Dad are both at the prime of life and brimming with joy. Also in the room is a glowing being who tells the scientist of an important project underway in which communication channels are starting to open up between "Heaven" and Earth. Her parents are to be involved in the project, along with a group of other spirit people, including some well-known scientists, some who died recently, others who died many years ago. They would like the lady scientist to be involved in the project from the Earth side because she happens to have the right psycho-spiritual qualities, which include a focused will that can penetrate the veil like an emotional lance and create realities in finer realms.

It all seems surreal, as dreams do, but the scientist says yes, she'd be willing to work on that project.

The next morning she awakens with only a vague memory of having had a dream that seemed to be important. So she relaxes her mind and lets it wander as it slows down to around 10 cycles per second, and suddenly the whole dream of her encounter with her parents and the angel and the talk of spirit scientists and an other-worldly communication project...it all comes back in a flash! And she realizes, it's her life mission! She springs out of bed and feels an unbridled exuberance as she steps into the new day...and into the rest of her life. She knows that her colleagues might question her sanity when she shares her experience with them, they try to talk her out of it, and she insists on its validity. She knows that she might be ostracized from the scientific community. But, you know, she doesn't care. That's not important to her. She's seen a bigger reality and chooses to be a part of it.

That, of course, is a completely fictitious story, but from my years of spiritual research I know for certain that it is a likely scenario of how the marriage of science and spirit could come about in the near future. I have colleagues who have gone through very similar transformations. I went through one myself. Now it's time for science to undergo such a transformation...led, perhaps, by a few very special, open-minded, open-hearted scientists!

PART FOUR

Resetting Our Inner Clock

I'M EXCITED TO SHARE a few things that I've learned over the years about achieving a good, resonant vibration, something that applies to many areas of our lives. An example of resonance in wide use today is a crystal oscillator, like the one shown here—a small quartz crystal encased in a hermetically sealed package. This one is a 4-megahertz oscillator, meaning the crystal inside vibrates four million times a second. Oscillators like this one act as a clock in PCs and laptop computers, providing a steady, reliable vibration around which the electronic circuitry performs.

Every crystal has a particular vibration. In fact, every material thing vibrates at a specific rate. Planet Earth has a resonant vibration, as does every rock and tree.

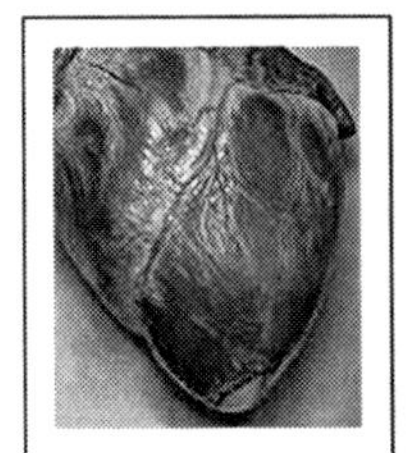

We humans have several important oscillators inside us. A biological oscillator, the heart, pumps blood at a variable rate. It's a very dense, slow oscillator that we can perceive with our five senses. *Lub-dub, lub-dub, lub-dub....* Our physical body as a whole is also an oscillator. Like all material things, it vibrates at a unique rhythm.

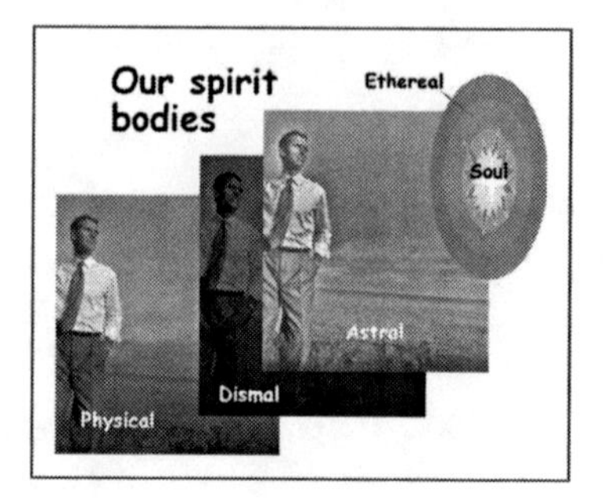

We human beings are also composed of several spiritual oscillators. The soul, in today's technical terms, could be called an infinite-megahertz oscillator. It consists of the purest spiritual light—a timeless, non-vibrating light direct from the source—from God. If you had to try to sum up the nature of that perfect light and consciousness in one familiar everyday word, that word would probably be "love." At the soul level we resonate with the soul of every other living thing everywhere. We are one with all, in a boundless ocean of pure love.

Between the soul and the physical body we have some ethereal and astral bodies inside us, and each of *them* vibrates at a different rhythm.

The key to spiritual mastery on Earth has always been the ability to bring our physical body into resonance with our finest spiritual bodies. One way to do that is to center our conscious, waking lives around love. And *that* can be done through such spiritual practices as prayer and meditation. Exercise and affirmations can also lend a hand in resonating at a love vibration. Keeping our body well-tuned by exercise and our mind well-tuned by affirmations can make this lifetime on Earth a happy, exciting adventure.

I'd like to end this book, then, with a couple of techniques to foster coherence of body, mind, and spirit. They offer a touch of yoga, a touch of isometric body-building, and a touch of meditation which, together, can help us reset our biological clock to resonate in love with our other, finer clocks...as we set sail toward paradise.

The techniques introduced in the coming pages include a heart meditation and a set of what I call mantric exercises. Yoga is one of the best regimens for personal development—centered as it is around meditation and exercise, but it's a major time commitment. If you have the time for yoga, by all means, practice yoga!

My techniques are simple and brief. The mantric exercises take about 20 minutes every morning and put you in a blissful mood for the day. They include not just five time-proven yoga exercises to tone and stretch the muscles and bones, but also some isometrics for body-building. Each exercise is done while reciting a mantra, which I've patterned after the Dalai Lama's daily affirmations mentioned in Part Two. After a few weeks of these mantric exercises, the world begins to look like a happier, less hostile place. After a few months, when you've learned to move smoothly to your heart during the meditation, you'll find yourself more peaceful and relaxed in all areas of your life. After a year of exercising, your body will look and feel better than it's looked and felt in years.

The meditation can take a few minutes or a half-hour—however long you wish—once or twice a day whenever you need to recharge or to calm down. Once you start entering a meditative state on a regular basis, your higher self will give you some guidance as to when and how long to meditate. Eventually, as your body, mind and spirit integrate more fully, you may be guided to make big lifestyle changes, especially nutrition—eating lighter foods, drinking more water, and taking vitamin supplements. If you've never meditated before, be prepared for paradise. The world will begin to look more vibrant. You'll see more colors than before. You'll find mystery and wonder as you gaze into the trees beyond the rushing stream. Beautiful music will touch you more deeply than it has before. You might catch a passing fragrance that stirs in you a memory of a familiar paradise existence from another time, another place, but profoundly real!

CHAPTER 21

Heart Meditation

FOR CENTURIES MYSTICS HAVE called the heart the seat of the soul, and when we move our awareness—our self—to the heart, we can feel an especially close connection to the source, to God. After all, the soul—the real you and me—is a piece of God. Here's a short script of a heart meditation I've used for years:

"As we sit or lie in a comfortable position, we close our eyes, and relax our body. We locate our awareness, usually behind the eyes, and move it slowly to the back of the head, then down the spine to the heart. It's not like we're in the head thinking about the heart; it feels as though we have moved our thinking process into the chest. The soul is the real you and me, a piece of the source, or God. Now, from this point at the center of our being, we absorb all the light, love and wisdom that we can possibly absorb from the source, from God. We feel our self, our soul, swell like a sponge. And as we swell with this divine light, we release it out into our spirit bodies and out further into our physical body. We let the light swirl through us, cleansing, purifying, and recharging us at every level. We pull more of this light from the source into our soul. As it swirls through us, we let it out into our surroundings, transforming this room into a sacred space, a temple of love and light. Now, as we remain centered in the heart, we think of our loved ones on Earth and in spirit, we think of our Ethereal guides, and we feel the close heart connection with all of these loving beings. We send streams of love from our heart to theirs, and we feel their love streaming into us."

Shortly after discovering the powerful effects of this meditation, I recited it slowly into a tape recorder as a guided visualization, leaving ten or twenty seconds of silence between sentences. I also added some nice, soothing music through the entire script. If you're interested, you're welcome to do the same. Feel free to tailor the script to your own beliefs and needs. If you're a Christian, it's wonderful to add Jesus to the meditation. If you're Muslim or Jewish or Buddhist or Hindu...or even *agnostic*, tailor the script to fit your beliefs. Having been an agnostic most of my life, the first thing I would have done during those years would be to change the word "God" to "higher power," or even remove it. Whatever works for you!

As you do the heart meditations everyday, the finer side of the supernatural begins to seem more natural. Angels and other loving, supportive spiritual influences move in close to provide assistance, protection, and guidance, because they all know the importance of spiritual work on Earth, and they are eager to help us enlighten our world. Know them by the warm, blissful feelings you get when they arrive. Welcome them. Encourage them. Communing with these finer beings is a subtle experience of insights, synchronicities, and an ever-growing zest for life.

You might want to get a copy of my 45-minute *Bridge to Paradise* CD. Not only can it accelerate spiritual development; it can also help prepare us for that final journey from *terra firma*–that one-way ticket to paradise at life's end–by raising our spiritual vibration.

I produced *Bridge to Paradise* in collaboration with Monroe Products of Lovingston VA. Our CD employs The Monroe Institute's *Hemi-Sync®* technology which automatically moves the brain into a meditative state, opening our mind to spiritual development. Listeners of the CD are guided through a heart meditation, then on to a journey to paradise, where we meet some Ethereal Beings.

Doing heart meditations regularly will help you remain clear, calm, and revitalized. That's enough for some people, while others find themselves drawn to the next level. Eventually you might feel pulled to apply your wisdom and skills in any of a variety of ways.

The applied heart meditations below are like other psycho-spiritual skills; although they can be developed to some degree by anyone, certain people have certain innate talents. Purifying the home works especially well for me, whereas spiritual counseling is a bit of a struggle.

As you meditate, be open to opportunities. The Ethereal guides working with you will know your strengths and weaknesses, and they'll come up with clever ways to guide you into situations that could benefit from your unique talents and skills. Although I'm referring here mostly to the skills you develop as a meditator, the same applies to your life skills and talents; you'll find doors opening for you more and more frequently as long as you don't ignore these opportunities. The more you embrace opportunities, the more they come your way! The less you heed them, the less they appear in your life. So, let's look at some of the many opportunities that might open up for you as you become comfortable with heart meditations.

Spiritual Purification of the Home

Become a conduit of this working light, and use the meditations at home to purify your house or apartment. After absorbing the light into your heart, and after purifying and revitalizing yourself at all levels, begin streaming it outward into your home, filling one room at a time, cleansing and purifying every corner of every room. Spend some minutes letting the light stream into and fill the entire home. This is all done in silence or with soft music playing in the background, using the mind to focus and direct the light stream.

If there's one room needing particular attention, maybe where a teen is living, spend extra time streaming light there, making it clear in your mind and in your silent announcement to the universe that there's no place in the home for confused or troubled spiritual energies. Only loving, supportive spiritual influences.

Purifying Other People's Homes

If you find you have the skill for this purification technique—if it seems to come easily and naturally to you—try cleansing other people's homes. Think of people of good will who might be in a difficult situation in life, and cleanse their homes the way you cleanse your own. It only takes a few minutes!

Don't send light to people who have chosen a dark, troubled path without regret. You'll be supporting or enabling their bad choices. You can ask Ethereal beings to help those troubled people, but it's better if you yourself focus on those who simply have made mistakes or are having problems and who need a boost to get where they want to be—in a happy place.

Cleanse the homes of the decent people the same way you cleansed your own home in the previous exercise. Stream light throughout their home during a heart meditation. It helps to be familiar with the layout of their home, but if you're not, you can simply stream light into the living space surrounding the person or family.

Rescue Work

Work with a spirit group to help lost and confused souls move to the light. Again, don't try to help malevolent spirits who've chosen darkness and chaos. Focus on lost souls who need help getting unstuck. This is something that will probably begin spontaneously. In a typical scenario, during a lucid dream, moments before you awaken, you might see a person in spirit (or several, or even a small crowd). They're looking at you, maybe with some confusion or maybe with some expectancy (or even a touch of desperation) in their eyes. Behind them you'll see light, as though they have their backs to a sunrise. You walk to them, and as you reach them, they turn around, and you watch them move to the light. Very simple, very quick.

Your guides and theirs are probably choreographing this meeting, because it often seems to take a person on Earth to facilitate these rescues. In many cases the spirit guides seem to have energies too fine and subtle to affect some spirits who are stuck in the dark, dismal realm and unable to see the light. Our dense earthly thoughts, focused through a love vibration, can help bridge the gap and bring the light into view.

Spiritual Counseling

With your honed instincts you'll be able to sense when someone you know is troubled, and why. It could be a friend or family member. You might simply mention that they seem a little down or troubled today, and see how they reply.

It might even be someone you know only casually at work or at church, or a member of a club, in which case you might broach the subject carefully. Don't do it unless you feel guided to do it, in which case the appropriate tone of voice and words will come to you. As you master this technique, your guides will work with their guides very quickly to choreograph these encounters.

Meanwhile, it's important to follow your inner voice when you encounter someone who's troubled. If you feel a warm compassion, it's probably a good idea to say something tactful. But if you get a dark or edgy reaction, move on.

If spiritual counseling turns out to be an innate talent of yours, word will spread, and you might even attract a growing clientele.

Teaching Others

When the student is ready, the teacher appears. You might be that teacher. If someone sees you meditating or notices your calm, happy demeanor and comments on it, don't hesitate to share a brief, general description of your inner work. Some people will insist on hearing more. Without being pushy, happily satisfy their curiosity.

For others, it will be obvious that they have no interest, and you can pleasantly walk away. No harm done.

A few people with dark clouds over their heads will invariably try to say something to wipe that silly smile off your face. The savage side shows up at surprising times in surprising ways. Go to your heart, let the hostility wash over you without taking it in, be calm, and move on.

But whenever a real student appears, feel free to teach. You might have a real knack for it!

So these are a few of the ways you can apply the skills that develop within you during your heart meditations.

As you meditate on a regular basis and remain centered in a good frame of mind with the intent to serve, you attract many fine, supportive loving spirits who will quickly identify your own particular psycho-spiritual strengths, and they'll help you foster them to become of greatest possible benefit to you and to the world.

Each person is unique in this way. There are countless beautiful, wonderful, and exciting spiritual techniques to apply to our world. Through dedicated practice you'll attract a team of competent spirit beings to help you put your innate spiritual skills to best use for yourself and for the world. This is the *real* way to be the best that you can be!

CHAPTER 22

Mantric Exercises

THESE MANTRIC EXERCISES STRENGTHEN the body-mind-spirit and set the groundwork for a happy life! They're based on the *Five Tibetan* yogic rites (which are believed to have been around for some 2,500 years) and *isometrics* (which are controversial), and they integrate mantras (affirmations) that lift the spirit. There's also controlled breathing throughout all the exercises. There are 12 exercise routines done in sets of five. They take about 20 minutes and are best done as close to daily as possible.

If you decide to do these exercises, as always, it's good to follow your inner voice about how often and how intensively to do them. What I offer are only suggestions. If you have any heart, lung, or blood pressure conditions, or any muscle irregularities, it's ***real important*** to get your physician's approval. Otherwise...enjoy!

About the isometrics: They build and tone the muscles, but they don't produce overall strength. If your aim in life is along the lines of arm-wrestling tournaments or stopping locomotives with your bare hands ☺, and if you have the time and inclination for long, grueling workouts, then you might prefer weightlifting. If you have the time and dedication for pure yoga, that's even better! But if your future involves normal activity—a job, a family, and no lifting of automobiles—and you want toned muscles and ligaments with minimal effort, isometrics are a good option. I like isometrics because they're fast and painless and can be done anywhere.

Between the two types of exercises—yogic (y) and isometric (i)—your heart and lungs get only a modest workout, so doing some swimming or hiking or biking or running from time to time can round things out.

Although I assume here that most people will do the routines first thing in the morning while lying on the floor, the isometrics (exercises 4-10) can easily be done almost anytime, anywhere—at a desk, in an airport terminal, and so on—without being too conspicuous. Instead of lying down, you can do some while sitting, others while standing.

Caution: During isometrics, flex gently the first few weeks to avoid strain...and keep breathing. Don't hold your breath.

Now let's try the mantras. Each is in 5 beats. Go ahead and recite the following mantras while breathing slowly, as indicated. (The underlines indicate accented syllables that come on the beat.) I like to recite them silently in my head so that spoken words don't get in the way of my breathing. If you prefer trying to talk while breathing, go for it! Silently (or aloud if you prefer) recite each of the three mantras below a few times, just to get a feel for them. Notice that for the yogic exercises (black figures) you exhale and inhale with every beat, while you cut your breathing in half during the isometrics (gray figures)—exhaling on one beat, inhaling on the next.

Beats (breaths)	1 (out-in)	2 (out-in)	3 (out-in)	4 (out-in)	5 (out-in)
Mantra 1 (y)	We're all	one, We	all want	love and	happiness.

Beats (breaths)	1 (out)	2 (in)	3 (out)	4 (in)	5 (out)
Mantra 2 (i)	I cherish	myself,	cherish	myself	today.
Beats (breaths)	6 (in)	7 (out)	8 (in)	9 (out)	10 (in)
Mantra 3 (i)	I cherish	others,	cherish	others	today.

The simple mantras contain important spiritual messages. By incorporating them into the exercises we plant the messages in our cells and tissues, deep within our carnal self, strengthening our noble side in a big way.

Exercises—Week One

Now let's combine the exercises and mantras. **Suggestion:** Do the following exercises everyday, or close to everyday, during the first week. If you're overweight or otherwise challenged in certain physical motions, adjust the exercises to something comfortable in style and amount.

Some of the isometric exercises concentrate on two different sets of muscles at the same time. If it seems difficult at first, it gets easier as weeks pass. If certain muscles, for example the calves, don't feel they're getting a good workout at first, just do the best you can and don't worry about it; the muscles will gradually get used to the exercises and will start to "kick in," so to speak.

1 – Big stretch (y): Stand up straight, hands dropped to your side, shoulders relaxed. Move your hands behind you and grasp the first two fingers of one hand with the other hand. Then do the following exercise routine while reciting the mantra below…

Beats (breaths)	**1 (out-in)**	**2 (out-in)**	**3 (out-in)**	**4 (out-in)**	**5 (out-in)**
Think **this**	We're all	one, We	all want	love and	happiness.
while doing ***this…***	*Chest out, head back, push hands back.*		*Bend forward, relax arms and head, touch the floor with your fingertips.*		

2 – Fast-breath spinner (y): Stand up, arms outstretched, then…

Beats (breaths)	**1 (out-in)**	**2 (out-in)**	**3 (out-in)**	**4 (out-in)**	**5 (out-in)**
Think **this**	We're all	one, We	all want	love and	happiness.
while doing ***this…***	*Spin 5 times to the right, 1 full circle every beat. Caution: If you feel dizzy, breathe more slowly or more shallowly.*				

3 – Leg lifts (y): Lie on your back, legs straight, hands beside you or under your tush, palms down, then…

Beats (breaths)	1 (out-in)	2 (out-in)	3 (out-in)	4 (out-in)	5 (out-in)
Think **this**	We're all	one, We	all want	love and	happiness.
while doing ***this…***	*Raise your feet above your head, then lower them to the ground. Lift and lower your head at the same time. Do that 5 times, once per beat.*				

4 – Leg-bicep, shoulder flex (i): Stay on your back and raise the knees so that the feet are flat on the floor. Stretch your arms across the floor straight out from the body, then…

Beats (breaths)	1 (out)	2 (in)	3 (out)	4 (in)	5 (out)
Think **this**	I cherish	myself	cherish	myself	today
while doing ***this…***	*Flex: 1) the muscles in the shoulders and 2) the leg biceps. Hold it for the duration of the mantra.*				

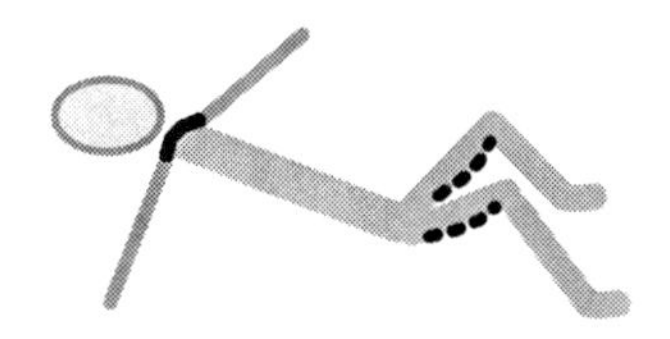

5 – Lap, shoulder flex (i): Still on your back, extend your legs together on the floor, put your hands on top of your head, then…

Beats (breaths)	1 (out)	2 (in)	3 (out)	4 (in)	5 (out)
Think **this** while doing ***this…***	I cherish	myself	cherish	myself	today
	Flex: 1) the shoulder and neck muscles, and 2) the lap muscles (front of the upper legs). Hold it for the duration of the mantra.				

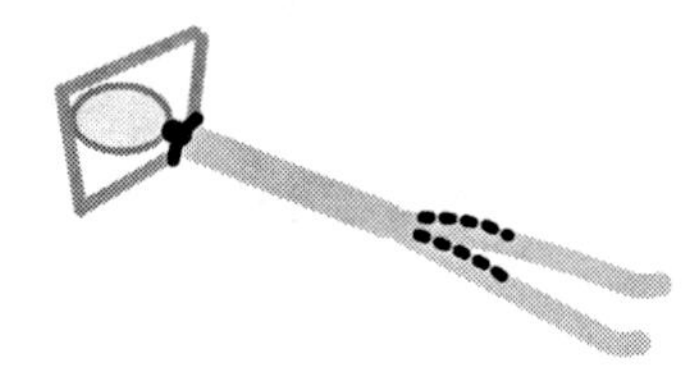

6 – Bicep, calf flex (i): Still on your back, raise your knees again with your feet flat on the floor, bend your elbows, then…

Beats (breaths)	1 (out)	2 (in)	3 (out)	4 (in)	5 (out)
Think **this** while doing ***this…***	I cherish	myself	cherish	myself	today
	Flex: 1) the triceps (back of the upper arms) and 2) the tush. Hold it for the duration of the mantra.				

7 – Tricep, tush flex (i): Still on your back, extend your legs and relax your arms on the floor again, then…

Beats (breaths)	1 (out)	2 (in)	3 (out)	4 (in)	5 (out)
Think **this** while doing ***this…***	I cherish	myself	cherish	myself	today
	Flex: 1) the triceps (back of the upper arms) and 2) the tush. Hold it for the duration of the mantra.				

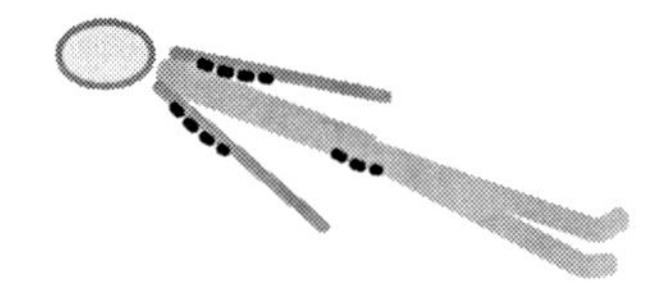

8 – Forearm flex (i): Still on your back…

Beats (breaths)	1 (out)	2 (in)	3 (out)	4 (in)	5 (out)
Think **this** while doing ***this…***	I cherish	myself	cherish	myself	today
	Bend the wrists forward, fists clenched.		*Bend the wrists backward, fists still clinched. Feel the forearms flex.*		

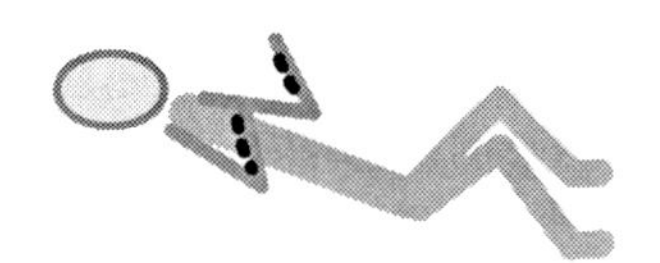

9 – Neck, shoulder, lower-calf flex (i): On your back, knees up and feet on the floor, clasp your hands above the groin, then…

Beats (breaths)	**1 (out)**	**2 (in)**	**3 (out)**	**4 (in)**	**5 (out)**
Think **this** while doing ***this…***	I cherish	myself	cherish	myself	today
	Flex: 1) Shoulders and neck, and 2) the small area between the ankles and calves. Raise your back off the floor so the weight is on your head, tush, and feet. Hold it for the duration of the mantra.				

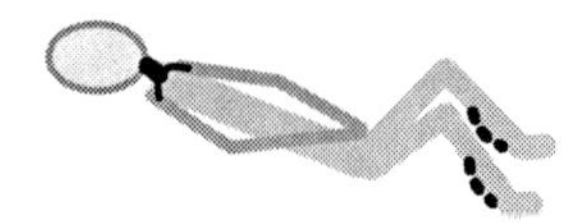

10 – Back-arch, belly, ankle flex (i): Lying on your back, put your hands palms down under your tush. Then…

Beats (breaths)	**1 (out)**	**2 (in)**	**3 (out)**	**4 (in)**	**5 (out)**
Think **this** while doing ***this…***	I cherish	myself	cherish	myself	today
	Arch your back and lift your feet six inches off the floor so that only your head, forearms, hands, and tush are touching the floor. Feel the pressure in the back and stomach while flexing your ankles. Hold it for the duration of the mantra.				

11 – Crab (y): Sit up with your legs together on the floor in front of you, put your palms down on the floor beside you, and look down at your navel. Then…

Beats (breaths)	1 (out-in)	2 (out-in)	3 (out-in)	4 (out-in)	5 (out-in)
Think **this** while doing ***this…***	We're all	one, We	all want	love and	happiness.
	Raise your torso toward the sky and bend your head back so you're looking behind you. Then return to the sitting position looking down. Do that 5 times, once per beat.				

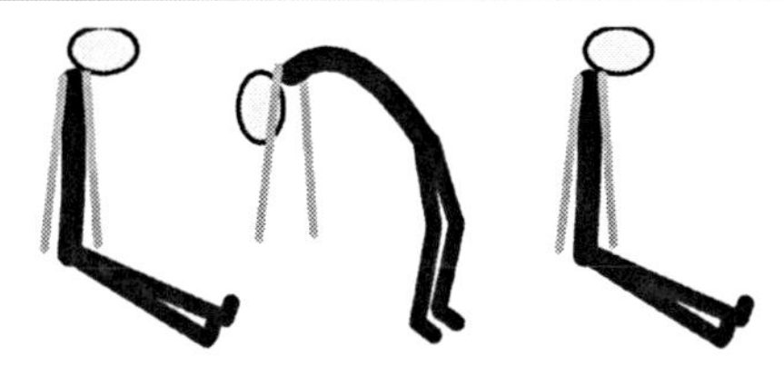

12 – Straight-arm pushups (y): Roll over onto your stomach. With your hands palms-down next to your chest, straighten your arms, pushing your torso up off the ground, pushing your tush straight into the air so your body's an inverted "V," and look back under your body. Then…

Beats (breaths)	1 (out-in)	2 (out-in)	3 (out-in)	4 (out-in)	5 (out-in)
Think **this** while doing ***this…***	We're all	one, We	all want	love and	happiness.
	Keeping your arms straight, lower your body to the "cobra" position with knees and toes touching the floor and your eyes looking up at the ceiling, then return to the inverted "V" position. Do that 5 times, once per beat.				

Exercises—Week Two

Now we'll double up. Again, do the following exercises everyday, or close to everyday, adjusting the style and repetitions according to your weight and other physical characteristics.

Exercises 1 through 3: Do each set twice.

Exercises 4 through 7: Do each set twice, but the second time through change the mantra to, "I cherish others, cherish others today," as shown below in the table for Exercise 8.

Exercise 8 (I): The second time through, change not just the mantra, but also open your hands and extend the fingers, as described...

Beats (breaths)	**1 (out)**	**2 (in)**	**3 (out)**	**4 (in)**	**5 (out)**
Think **this** while doing ***this...***	I cherish	myself,	cherish	myself	today
	Bend the wrists forward, fists clenched.		*Bend the wrists backward with fists still clenched. Feel the forearms flex.*		
Beats (breaths)	**6 (in)**	**7 (out)**	**8 (in)**	**9 (out)**	**10 (in)**
Think **this** while doing ***this...***	I cherish	others,	cherish	others	today.
	Bend the wrists forward, ***hands open****.*		*Bend the wrists backward,* ***hands open****. Feel the forearms flex.*		

Exercises 9 and 10: Do each set twice, but the second time through change the mantra to, "I cherish others, cherish others today," as with the other isometrics above.

Exercises 11 and 12: Do each set twice.

Exercises—Week Three

Now we'll add another set. (Again, adjust the routines according to your physique.)

Exercises 1 through 3: Do each set three times.

Exercises 4 through 10: For each exercise, follow the regimen for Week Two then the regimen for Week One.

Exercises 11 and 12: Do each set three times.

Exercises—Week Four

Now we'll add the fourth set. (Again, adjust as necessary.)

Exercises 1 through 3: Do each set four times.

xercises 4 through 10: For each exercise, follow the regimen for Week Two twice.

Exercises 11 and 12 Do each set four times.

Exercises—Week Five and Beyond

Now we'll add the fifth set. (Adjust the routines as necessary.)

Exercises 1 through 3: Do each set five times.

Exercises 4 through 10: For each exercise, follow the regimen for Week Two twice, then follow the regimen for Week One.

Exercises 11 and 12: Do each set five times.

Doing these exercises daily, or at least several times a week, will keep your body fit, while the mantras will program your mind to be blissful. Do these mantric exercises regularly, and you'll have no choice but to be fit and happy.

Note: The human body comes in a wide assortment of shapes and sizes. If these exercises don't quite work for you, feel free to tailor them in a way that does work. For a complete exercise system, do some occasional walking, running, biking, dancing, swimming...whatever aerobics you enjoy.

Come to think of it, today's nations, religions, multinational corporations, and other macro systems also come in a wide assortment of shapes and sizes. I'd love to see the United Nations given the authority to try to help those macro systems the same way this book is trying to help you with these exercises. An empowered UN wouldn't issue strict rules and rigid madates. Rather, it would issue advisories to macro systems: "Based on this information...we give you these recommendations...." Then it would be up to the macro systems to accept or reject the UN advisories. Both the advisories and the compliance or noncompliance would become public record. That could be the ideal way to initiate world-level regulation through the UN!

Exercises—Quick-Reference Guide

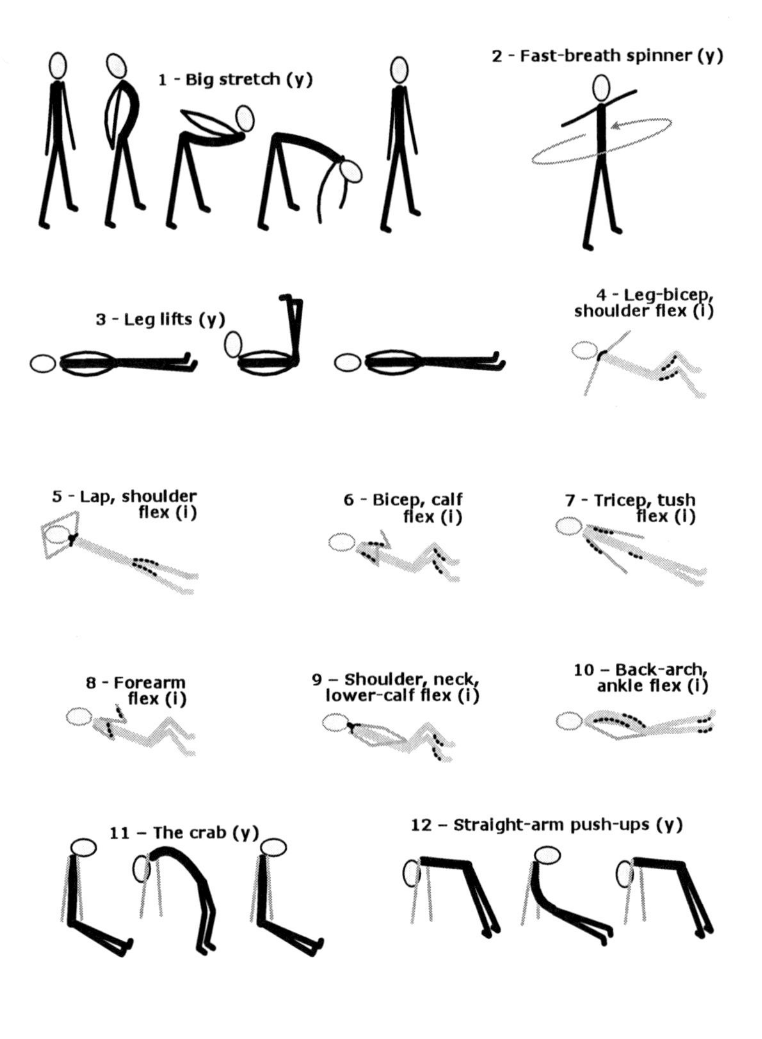

CHAPTER 23

Our Noble Destiny

WHEN ETHEREAL BEINGS COME close to our world to observe humanity, I suspect it's like a walk in the garden for them. Mothers' love for their children is spread everywhere like beautiful roses. People's passion and commitment to protect the less fortunate are like hearty evergreen shrubs. Prosperous communities of people working together on compatible projects are like bushes laden with berries. But when Ethereal beings look under rocks in the garden, all sorts of repulsive things are crawling and slithering around, trying to escape the light. People caught up in lying, cheating, stealing, murdering, fear-mongering, war-mongering, and character assassination crawl and slither in the shadows.[43] There's constant tension in the garden between the noble side embracing the light and the savage side groveling in the dark.

The Ethereals want to help us all find the light. They work with decent people of good intent, and they comfort those who are suffering. But there's only so much they can do. Through ITC systems the Ethereal beings have told us at different times in various ways that humanity today is approaching an End Time, and our fate rests mostly on our own collective shoulders. In the Fall of 1995 they said:

Much depends on you and on your decisions those days....

In the following paragraphs I'll paraphrase what might be the most important message[44] ever brought into our world during this Epoch—maybe any Epoch—in which they gave us important hints about why we are (our purpose...) whom we are (as human beings...) where we are (in societies within ecosystems on Earth...), and what's in store for us if we make the right choices:

Project Sothis began long ago, before Atlantis, and was orchestrated by finer spiritual forces. It has been an effort to establish on Earth a doorway to spiritual realms, not just for communication with those realms but to transport us back and forth. To achieve that goal, those spiritual forces developed a race of humans savage enough to tame a merciless terrestrial environment, and noble enough to turn it into a thriving civilization—a paradise. Our savage side still dominates much of human affairs, and so nonphysical beings from many realms have taken an interest in humanity in the Second Epoch—for better or worse—to determine the fate of the Project. There are Ethereal beings who support our noble side in an effort to bring The Project to fruition, and there are dismal spirits who stir up our savage side and work toward our destruction.

The Ethereals want to forge a lasting relationship with us, but we have to be able to rise to the occasion. It won't be easy because of our own savage nature and the dismal spirits who stir up all sorts of problems (fear, greed, animosity, conflicts...). The Ethereal beings can reduce the effects of this dark storm, but they can't prevent it.

While there are countless brilliant beings in the ethereal realms, a group of seven Ethereals told us they are assigned to provide guidance and support for the Project. They'll work with people who commit to a moral course, which involves discerning information coming from finer realms, acknowledging it, adjusting it to fit our world, and acting on it. That's not the same as religion, which requires faith. Faith is one thing, the proactive moral course something else. Establishing an ethereal partnership involves the ability to resonate with and to be open to finer spiritual influences.

The Seven Ethereals have chosen certain cultures or civilizations six times in the past through which to extend a lifeline to humanity so that we might rise as noble humans to continue The Project. Those ancient times of miracles were the subject of legend, myth, and inspired religious texts. Today those Seven Ethereal beings are back, and the chosen people, I believe, are going to be all of humankind, not just one isolated culture as in the past. They want to establish and sustain contact with an association or network of people representing all humankind.

That, in a great, big nutshell, is what the Ethereals told us in that message and other messages: They want to develop a lasting relationship with humankind, but we need to avoid the pitfalls stirred up by our savage side as we rise to the occasion.

Most of us at some time in our lives have cherished, trusted, and felt unbounded good will toward certain people and pets. That's the noble side of humanity that gives our spirit a golden glow, and that's the part of our nature that holds in the balance our salvation, our destiny—the completion of The Project.

During my years with INIT I could recognize the influence of the Ethereals by the light, purely positive feelings I got when they were involved with our group. There wasn't a shred of deception on their part nor uneasiness on mine—just pure, unadulterated love. Their support of our group faded away around the year 2000, but I'm certain they'll be back when conditions are right—that is, when we have learned to master our noble side.

The biggest responsibility of social systems in the future will be to establish conditions that promote those feelings of trust and good will within and among the people. There are political, economic, and scientific techniques and technologies that can move us in that direction, ranging from right regulation and ***E***-conometrics to brain implants, but the greatest force is within each of us. As individuals we can polish ourselves up with meditation, prayer, exercise, and other forms of inner work.

We all know what it feels like when we're inspired by the noble side or gripped by the savage side. With noble inspiration we might get a warm feeling in the heart. In the grips of the savage side there might be a tightness in the pit of the stomach. Being consumed by fear, animosity, resentment, and other savage emotions lowers the spiritual vibration, increasing the likelihood of our getting stuck in a dismal spirit realm for awhile after we die. The simple key to eternal joy has always been to live in love, trust, and good will as much as reasonably prudent while alive on Earth. Fostering the noble side raises our spiritual vibration, carrying us to paradise after we die.

As more people on Earth choose a noble path, the spiritual vibration of the planet rises too, attracting Ethereal beings into our world to manifest miracles and joy everywhere. This is the destiny I want for our world, and I hope this book helps to move us in that direction!

Index

ENDNOTES

[1] From: *Contact! a triannual report of technical spirit communication*, a journal I published between 1996 and 1999 to report the ITC communications received by our group, the International Network for Instrumental Transcommunication. (Back issues are on file at www.spiritfaces.com.) Issue 96/02, p.6. Continuing Life Research, 1996 May-August:

The two letters below from an ethereal being (angel) named Thfirrin were planted paranormally in the computer at INIT Station Luxembourg in 1995, on October 15 and October 20, respectively. At the time of those contacts we (INIT members) had already heard from three of The Seven ethereal beings. Technician (as experimenters had called him since 1986) had been facilitating and protecting ITC development for a decade. Ishkumar and Thfirrin then began making contact in 1994, shortly after a group of us had decided to have a founding meeting for INIT the following year. Before the following contacts were received, experimenters had been asking our spirit friends questions relating to Christian beliefs.

Letter # 1. *Long, long ago when humans came to Earth from Eden (or Marduk), they lost mastery over nature after their dissension with the serpent. This was the start of a drama which would cast its* *shadow even to the beginning of a new religion of a new God. Humans had to fight against nature and some of its most dangerous creatures. You call them monsters, but they were only life forms which defended themselves and their kind. These creatures did not subject themselves to the intellectual superiority of humans and had to be defeated and destroyed by cunning and strength. Human civilizations prevailed. Of these times long, long ago, only legends were handed down. Do not scorn them. The day may not be far off when your sciences will dig up the skeleton of a dragon.By the time Christianity started its conquest, Earth had been freed of these monstrous animals, but they were still very much present in people's minds. Now it was saints such as Michael who defeated them by force. Later on, others whom you call saints subdued the so-called wild animals up to the day when force became superfluous and they were subdued by intellect....*

Letter # 2: *At this point we wish to return to a subject of frequent discussions (especially among Catholic Christians): the subject of original sin. If we were to analyze and to accept verbatim what is still believed and frequently taught in Catechism today, which is the so-called "fall of Man" and its consequences, every logically thinking person, Catholic or not, would have to conclude that Man's punishment did not bear any relationship to his transgression. Because, curiously enough, the first sinful humans were said to have been forgiven and the burden of this sin placed on their (innocent) progeny. This entire, contradicting story was to serve as an explanation for death. Actually, the "fall of Man" is a legend handed down from Greek mythology. It was maintained by the "Mysteries" and found its way into the schools of philosophy during the Greek classic period. Humans were considered to be descendants of Titans who had killed the young Dionysus-Zagreus. The burden of this crime weighed* *heavily on them. In a writing by (early Greek scientist and philosopher) Anaximander it is said that the unity of the world was destroyed by a prehistoric crime. In reality, these legends are based on a factual incident: the downfall of the civilization you often call Atlantis (also known by other names). This downfall was brought about by the descendants of the last inhabitants of Marduk who became marooned on planet Terra. This downfall came through reliance on and blind trust in a massive technology. Later, the Church tried to prove the theory of the consequences of original sin by emphasizing Man's tendency toward evil. The Church's explanation was that Adam, who was free of original sin, had no inclination to evil (in my first text I allegorically referred to it as "the serpent"). You see, there was a "fall of Man," but different from what you imagined. We know that many among you will not believe us. This does not bother us. We know that this is how it was. Today you may consider it science fiction, but you too will one day, once more, recognize its validity. The next time we shall speak of more pleasant things.*

—Thfirrin, one of The Seven

[2] From: *The Holy Bible*; King James Version. Genesis 2, 8-24:

The LORD God planted a garden eastward in Eden; and there he put the man whom he had formed. And out of the ground made the LORD God to grow every tree that is pleasant to the sight, and good for food; the tree of life also in the midst of the garden, and the tree of knowledge of good and evil... And the LORD God took the man, and put him into the garden of Eden to dress it and to keep it. And the LORD God commanded the man, saying, "Of every tree of the garden thou mayest freely eat: but of the tree of the knowledge of good and evil, thou shalt not eat of it: for in the day that thou eatest thereof thou shalt surely die."

And the LORD God said, It is not good that the man should be alone; I will make a helper for him... And the LORD God caused a deep sleep to fall upon Adam, and he slept; and he took one of his ribs, and closed up the flesh instead thereof. And the rib, which the LORD God had taken from man, made he a woman, and brought her unto the man. And Adam said, This is now bone of my bones, and flesh of my flesh: she shall be called Woman, because she was taken out of Man. Therefore shall a man leave his father and his mother, and shall cleave unto his wife: and they shall be one flesh. And they were both naked, the man and his wife, and were not ashamed.

[3] Geodynamist John Huw Davies (Cardiff University in England) announced that the strange physical characteristics of planet Venus (including reverse spin and parched, scorched landscape) could be explained by an age-old head-on collision with another large body. His article was published in the January 31, 2008, issue of *Earth and Planetary Science Letters.*

[4] From: Nikola Tesla's Teleforce and Telegeodynamics Proposal, Leland I. Anderson, Editor. Twenty-First Century Books:

"These (vibrations fed into the Earth) are generated by Tele-Geo-Dynamic transmitters which are reciprocating engines of extreme simplicity adapted to impress isochronous vibrations upon the earth, thereby causing the propagation of corresponding rhythmical disturbances through the same which are, essentially, sound waves like those conveyed through the air and ether. . . . With a machine of this kind it will be practicable, in the differentiation of densities and aggregate states of subterranean strata and tracing their outlines on the earth's surface, to reach a precision approximating that which is secured in the investigation of the internal structure of bodies by penetrative rays. For just as the vacuum tube projects Roentgen shadows on a fluorescent screen, so the transmitter produces on the earth's surface shadows which can be detected by acoustic devices or rendered visible by optical instruments. The receiver can be made so sensitive that prospecting may be accomplished while riding in a car and without limit of distance from the transmitter."

[5] From *Contact!* 1997, Issue 01, p.14, message from spirit of Dr. Swejen Salter:

The inhabitants of Earth have not only a different view of their world than do those of my home planet, Varid, but also a different physics. For instance, the planet I come from was flat in 600 B.C. It was as flat then as it is round now. The planet has not changed, but the spirit of its inhabitants has. The spirit of man, whether on Earth or Varid, is constantly changing. Man attributes this maturing of intellect to advancements in science and to more accurate methods of investigation.

[6] From *Science Daily*, Jan. 15, 2008: Vanderbilt University Graduate student Maria Couppis conducted a project for her doctoral thesis in which the female mouse was removed from a male-female couple sharing a cage, and replaced by an intruder male mouse. The home male mouse became aggressive toward the intruder male until the intruder was removed from the cage.

A benign trigger was placed in the cage once a day with the lone male. Whenever he pushed it with his nose, an intruder male would be placed in the cage, and a brawl would ensue. Ms. Couppis placed the trigger in the cage once a day, and everyday the home male pushed the trigger to bring the intruder into the cage, suggesting that the home male mouse enjoyed the "reward" of a good brawl.

The results were consistent for other mice under the same conditions; they seemed to enjoy "bringing it on." Then when the mice's dopamine receptors were suppressed with drugs (effectively disabling the body's built-in reward drugs), the home male mice didn't push the trigger as much, indicating that a good scrap was no longer as rewarding.

Vanderbilt Special Ed professor Craig Kennedy concluded from his student's project that aggression is its own reward, as aggressive events activate the same "reward pathways" in the brain that are activated by food and sex. And it's long been known that the behavior of mice and men and other mammals is similar when the various species find themselves under similar conditions.

[7] Italian archaeologist Andrea Carandini has been conducting excavations for twenty years at the site of the Roman Forum and discovered in 2007 the remains of a 3,700-square-foot royal palace dating back to the time of the city's legendary beginning. The remains were found at the precise spot where the Temple of Romulus stands today. Said Rome's superintendent for monuments, Eugenio La Rocca, "Someone created the legend of Rome's founding with knowledge that it had a historic basis. The details handed down by most of the Latin writers is much more than speculation."

[8] Dean Falk of Florida State University and her colleagues named a newly discovered species of little people "Homo Floresiensis," after the East Asian island of Flores where their

12,000-year-old remains were found. The little people had long arms and sloping chin, suggesting a more primitive human than we. But sophisticated tools and evidence of a campfire were found with the remains, indicating a more advanced lifestyle than their anatomy would imply. There's a stir among scientists, some of whom feel this "Hobbit" clan represents a hitherto unknown species, while others argue that they were regular Cro-Magnons who suffered hypothyroidism. (It makes most sense to me, of course, that the "Hobbit" people were just one of many human varieties genetically engineered by the Edenites and their descendants here on Earth.)

[9] The photo below, taken at a London train depot, appeared in *Strand Magazine*, December 1895. The 12-foot-tall "fossilized Irish giant" was dug up in County Antrim by a Mr Dyer while prospecting for iron ore. It was exhibited in Dublin, Liverpool and Manchester.

[10] From the *Bible*, Genesis 6:

When men began to increase in number on the earth and daughters were born to them, the sons of God saw that the daughters of men were beautiful, and they married any of them they chose. Then the LORD said, "My Spirit will not contend with man forever, for he is mortal; his days will be a hundred and twenty years." The Nephilim were on the earth in those days—and also afterward—when the sons of God went to the daughters of men and had children by them. They were the heroes of old, men of renown. (More modern translations replace "sons of God" with "the divine beings.")

[11] From *Contact!* a triannual report of technical spirit communication. Issue 97/03, p.5. Continuing Life Research, 1997 September-December:

Members of INIT received several contacts referring to Project Sothis and Atlantis. One contact reported that Atlantis had been located near the modern-day island of Helgoland (located in the North Sea, right off the small German Atlantic coastline.) This information astonished many INIT members. Jules and Maggy Harsch who received the information had always assumed Atlantis was in the Adriatic Sea, off the coast of Greece, as many people today believe. I was uncertain as to whether Atlantis really existed, or was part of ancient mythology. My friend and colleague Hans Heckmann had worked in the 1980s with Dr. William Francis Gray Swann's spirit group, who had described Atlantis as originally stretching from the present Bermuda triangle, all the way north to the English Channel. Other people thought the ruins would be found in the Caribbean. So, there were many mixed views of Atlantis among INIT members, but most of us had grown to trust our spirit colleagues at Timestream and the information they were giving us, in the same way that we trusted people on Earth who spoke to us in reasonable terms consistently over a period of time. This is what they told us in a computer contact in

1997, received in the INIT computer at Station Luxembourg:

The project started 20,000 years ago. Its final phase took place in Atlantis. Basilaie was the king's island in an empire that sank in the middle of the 13th century B.C. It was located east of Helgoland. Diving expeditions have meanwhile located in the waters of a boulder field the ruins of manmade walls. Plato's report (the only document still in existence) corresponds with the cultural traces of the South Scandinavian and Danish Bronze age. There also is a parallel with the stories of natural catastrophes and devastation which took place around 1220 B.C. in Northern Europe. For a better understanding also read the work of pastor Juergen Spanuth, The Demystified Atlantis, published by him in 1953. The Atlanteans were not necessarily the kind of people that many of you imagine. They did have a great culture, but they also had a great fleet that attacked and robbed the Mediterranean countries (where did you think their legendary wealth came from?) among them Corsica, Crete. Greece, Thessalia, Macedonia etc. Also see the big Egyptian relief of a naval battle (the Egyptian priests reported to Solon about these raids—see Plato.)

See also in the temple of medinet-habu, armament of the warrior sailors such as long swords, horned helmets, ships' bows built like the neck of a swan....

We all wish you a Merry Christmas from Timestream Spirit Group!

[12] From: *Psychopharmacology*, January 13, 2008. "The rewarding effect of aggression is reduced by nucleus accumbens dopamine receptor antagonism in mice." Dopamine (DA) receptors within the nucleus accumbens (NAc) are implicated in the rewarding properties of stimuli. Aggressive behavior can be reinforcing but the involvement of NAc DA in the reinforcing effects of aggression has yet to be demonstrated.

[13] From: *Journal of Neuroscience*. August 15, 1999. "Induced Analgesia Mediated by Mesolimbic Reward Circuits." Rats chronically implanted with subcutaneous morphine pellets demonstrated tolerance to the antinociceptive effects of acute systemic morphine administration but did not show cross-tolerance to NSIA....

[14] From: http://opioids.com/dopamine/: Researchers have long known that the body can activate its own form of pain relief in response to painful stimuli. Now, UC San Francisco investigators have determined that, in rats, this long-lasting relief is produced by the brain's reward pathway...and the relief was as potent as a high dose of morphine. While various individual structures in the brain have been known to produce analgesia, or pain relief, when electrically stimulated or exposed to narcotic painkillers, the finding provides the first physiological evidence that pain itself elicits analgesia....

From: http://learn.genetics.utah.edu/units/addiction/drugs/: Within seconds of entering the body, drugs cause dramatic changes to synapses in the brain. By bypassing the five senses and directly activating the brain's reward circuitry fast and hard, drugs can cause a jolt of intense pleasure. As the brain continues to adapt to the presence of the drug, regions outside of the reward pathway are also affected. Brain regions responsible for judgment, learning and memory begin to physically change or become "hard-wired." Once this happens, drug-seeking behavior becomes driven by habit, almost reflex. This is how a drug user becomes transformed into a drug addict....

[15] From *Science*, October 2007: Nasir Neqvi at the University of Iowa found that strongly addicted cigarette smokers who

injure their insular cortex (from stroke or accident, for example) immediately lose their cravings for nicotine. Marco Contreras, Francisco Ceric, Fernando Torrealba found that temporarily disabling the insular cortex of amphetamine-conditioned rats virtually eliminates their immediate cravings for amphetamines.

[16] Robert Monroe founded The Monroe Institute and wrote such landmark books as *Journeys Out of the Body*, *Far Journeys*, and *Ultimate Journey*. Through his experiences he developed the Hemi-Sync method of brain hemispheric synchronization, which has helped hundreds of thousands of people journey beyond the physical world into subtle realms to explore spiritual realities. Thanks to Monroe and Hemi-Sync, they've learned to do in minutes what can take aspiring mystics years to achieve.

[17] From three websites:

http://www.webmd.com/stroke/news/20040415/brain-implants

http://www.sciact.org/articles/articlepage.asp?Pageid=202

http://world2come.tribe.net/thread/074fab78-5f97-4d42-b1d1-6c39d49315d4

Boston—Cyberkinetics Inc. of Foxboro, Mass., has received Food and Drug Administration approval to begin a clinical trial in which four-square-millimeter chips will be placed beneath the skulls of paralyzed patients....

[18] From: *Nature* 442, 164-171 (13 July 2006). "Neuronal ensemble control of prosthetic devices by a human with tetraplegia." Neuromotor prostheses (NMPs) aim to replace or restore lost motor functions in paralyzed humans by routing movement-related signals from the brain, around damaged parts of the nervous system, to external effectors. To translate preclinical results from intact animals to a clinically useful NMP,

movement signals must persist in cortex after spinal cord injury and be engaged by movement intent when sensory inputs and limb movement are long absent....

[19] From: www.iplant.eu

Dopamine and serotonin control fundamental functions in the brain, particularly in motivation, mood, learning and creativity. Implants that electrically regulate the release of these neurotransmitters have been used to produce specific brain states and behaviors like learning and physical exersice in rats. This site promotes the development of such brain implants, here called iPlants, for humans. iPlants would allow people to program their own behavior and mood.... (Christopher Harris, research assistant, University of Sussex, Brighton UK)

http://www.betterhumans.com/forums/thread/17570.aspx

Here's a brain implant that will let you program yourself: Let's define an iPlant as eight arrays of stimulating electrodes that give their user control of the release of dopamine and serotonin in his or her brain by regulating electrical activity in the VTA, SNc, dorsal and medial raphe nuclei. As you may know, dopamine determines what we consider important and rewarding, and thus what we feel motivated to do, whereas serotonin has a strong influence on our mood and our ability to re-evaluate.... (Kringelbach et al, 2007; Perlmutter & Mink, 2006) (Burgess et al, 1991) (Garner et al, 1991). (Talwar et al, 2002). http://www.iPlant.eu

[20] From two websites:

http://www.exceltreatment.com/research/: Excel Treatment Program, Denver, Colorado, USA http://detoxmanagement.com/research/3.html: Detox Management Group, Reno, Nevada, USA

About one-fourth of the human population have an A1 allele (gene) in their DNA, and the others have an A2 allele. Most treatments take into account the fact that the A1 allele is found in a larger proportion of addictive-compulsive personalities.

[21] From: *Behavioral Neuroscience*, October, 2004 (Vol. 118, No. 5). "Stress and Aggression Reinforce Each Other at the Biological Level, Creating a Vicious Cycle."

Behavioral neuroscientists in Amsterdam discovered a connection between hormones, stress, and aggression. They found that electrical stimulation of the aggression-control centers (hypothalamus) in the brains of 53 male rats caused more stress hormone (corticosterone, which is like human cortisol) to enter the blood, making the rats more violent and aggressive. Later they injected the hormone into the blood, producing the same aggressive behavior associated with the brain's aggression center. The scientists, led by Menno Kruk, PhD, of the Leiden/Amsterdam Center for Drug Research, concluded that stress and aggression feed on each other very quickly. The findings might help explain road rage during stressful traffic conditions, and why aggressive behavior and stress sometimes enter a feedback loop in human relationships that is hard to stop. It could explain why modern humanity today might be in a self-perpetuating cycle of violence manifesting in movies, on the battlefield, in the streets, and in families.

[22] "Probably the greatest threat from genetically altered crops is the insertion of modified virus and insect virus genes into crops. It has been shown in the laboratory that genetic

recombination will create highly virulent new viruses from such constructions. Certainly the widely used cauliflower mosaic virus is a potentially dangerous gene. It is a pararetrovirus meaning that it multiplies by making DNA from RNA messages. It is very similar to the Hepatitis B virus and related to HIV. Modified viruses could cause famine by destroying crops or cause human and animal diseases of tremendous power." Dr. Joseph Cummins, Professor Emeritus of Genetics, University of Western Ontario.

"Genetic engineering bypasses conventional breeding by using artificially constructed parasitic genetic elements, including viruses, as vectors to carry and smuggle genes into cells. Once inside cells, these vectors slot themselves into the host genome. The insertion of foreign genes into the host genome has long been known to have many harmful and fatal effects including cancer of the organism." Professor Mae Wan-Ho, Department of Biology, Open University, UK.

Most of the following references (unless otherwise noted) are chapters from my anthologies: *Solutions for a Troubled World* (Earthview Press, 1987) and *Healing the World…and Me* (Knowledge Systems, 1991).

[23] Majid Rahnema: "Swadhyaya, the Silent, Singing Revolution of India;" Ch.7, Healing the World… (Iran. Retired Persian ambassador and a grassroots learner-philosopher. Many articles and seminars on societal, transcultural, and international issues, focusing on impacts of modernization on poor cultures.)

[24] Jan Tinbergen: "Wise Management for a More Humane World;" Ch.5, Solutions for a Troubled World (Netherlands. Economist, Nobel Prize winner and author of the Report to the Club of Rome)

[25] Howard Richards: "World Values for Economic Justice;" Ch.19, Solutions for a Troubled World (USA. Professor of Peace Studies, Earlham College)

[26] John Fobes: "The Next Step Toward World Order;" Ch.7, Solutions for a Troubled World (USA. Original delegate to the UN Secretariat in 1945; international official, advisor to NATO, director-general of UNESCO administration...) based in New York, Washington, London, Paris, and New Delhi)

[27] Marc Nerfin: "Five Changes for the United Nations;" Ch.8, Solutions for a Troubled World (Switzerland. Journalist, teacher, and officer/member of many international forums and foundations. Has lived in Tunisia, Ethiopia, New York City, Mexico, and Rome, and has visited half of the UN member nations. Former editor of IFDA Dossier.)

[28] Hanna Newcomb: "Peace Values for a Better World;" Ch.9, Solutions for a Troubled World (Canada. Acclaimed, award-winning Czeck/Canadian peace activist, Quaker, and member and officer of many international groups, including the UN Association, the Voice of Women, the World Federalists of Canada, and the World Law Foundation.)

[29] Gerald Mische: "Redefining Sovereignty in an Interdependent World;" Ch.13, Solutions for a Troubled World (USA. Co-founder and former President of Global Education Associates with affiliates in more than 60 countries. Co-founder and first director of Association for International Development. Writer and lecturer who conducted world order workshops on five continents.)

[30] Keith Suter: "The Common Heritage of Humankind;" Ch.12 Healing the World... (Australia. Peace activist and

independent Christian Scholar; former director, Trinity Peace Research Institute)

[31] Hanna Newcomb (see above)

[32] John Chidley and Keith Clarke: "On the Road Toward World-wide Communication;" Ch.3, Solutions for a Troubled World (United Kingdom. Chidley: Doctor of Mathematics with expertise in the distribution of information through telephone networks; author of many articles on telecommunication standards. Clarke: Engineering background in computer time-sharing, computer graphics, computer-aided design, videotext, and telecommunication standards, on which he's written many articles.)

[33] Prachoomsuk Achava-Amrung: "Communication Education;" Ch.2, Solutions for a Troubled World (Thailand; Ed.D. in Economics of Education and Research Methodology; peace activist, headed many research projects and education organizations at the university level; author, editor, and translator of many books and articles.)

[34] Jan van der Linden: "The Way of Meditation;" Ch.1 Solutions for a Troubled World (Netherlands. Late director, School for Esoteric Studies, NYC; Dutch ascetic studying metaphysics since 1945.)

[35] Patricia Mische: "Global Spirituality;" Ch.15 Solutions for a Troubled World (USA. Co-founded Global Education Associates with affiliates in more than 60 countries and conducted more than 1,000 workshops around the world on peace, justice, world order, and other global concerns.)

[36] Robert Muller: "Global Healing through Global Education;" Ch.10 Healing the World… (France and Germany. During 38 years of service to the UN, held many positions, including assistant

to three Secretaries-General, and earned many nicknames, including UN Prophet, Optimist-in-Residence, and First 21st-Century Man. Many awards for global service and humanism.)

[37] Hilka Pietila: "Beyond the Brundtland Report;" Ch.13 Healing the World... (Finland. Secretary-General Finnish UN Association. Member and official in many national and international groups, including Women for Mutual Security, Women for Peace, and International Foundation for Development Alternatives [IFDA])

[38] Liang Jimin and Wang Xiangying: "Family Planning in China;" Ch.5 Healing the World... (China. Liang: M.D. and Vice President, Family Planning Association; teaches Medicine and Population Studies at Hebei University. Many administrative posts in associations dealing with population, demographics, and eugenics. Wang: Division chief of the State Family Planning Commission; involved with many international seminars, committees, conferences, articles and books, mostly on family planning, economics, and international relations.)

[39] July 2008. *National Geographic*. "Who Killed the Virunga Gorillas?" by Mark Jenkins (photos by Brent Stirton).

July 6, 2008. *60 Minutes*, "Kings of Congo," produced by Robert G. Anderson and Casey Morgan.

[40] Ahmad Abubakar: "Steps to a Stable World Economy;" Ch.17, Solutions for a Troubled World (Tanzania. Bs.C. economics. Educated in the East [International Institute of Management, Romania], in the West [Vanderbilt University, USA], and in the South [University of Ibadan, Nigeria], and pursued study and research in Romania, USA, Egypt, India, Malaysia, Japan, Brazil, and throughout Africa.)

[41] JS Mathur: "New Values for Equitable Growth;" Ch.18, Solutions for a Troubled World (India. Director of the Institute of Gandhian Thought and Peace Studies; writer, editor, professor, and international lecturer on social, industrial, economic, and peace issues. Author of dozens of articles and 20 books.)

[42] Howard Richards (see above)

[43] Computer contact by Ishkumar, one of The Seven, Station Luxembourg, 1996 July 19, 11:20 a.m.:

Children of Earth, people of Terra. You know the world is not changed by cosmic events but by changes in the individual. Every person is unique and he can build a palace for good or a dungeon for evil. This recognition should make clear to all of you your share of responsibility in all happenings. Especially in ITC it is important that everyone should be aware that he or she can play a decisive part. The spirit of time was favorable for you people. The positive reform first coming from a few individuals has now manifested in a group which is forming around you. It is only natural that some of them who want to hold on to what was overtaken are hostile toward you. They want to try to suppress the spring that is now bubbling up, so it sinks back into the ground. They will not succeed. They have once before experienced that the subterranean stream of water that developed appeared anew with even greater power. Some of you think every human being carries a spark of something higher within and believe they can argue with everybody on the same level. Unfortunately, this is not true. It is easy to forget that you are still on the physical side of the veil and have your daily battles with all the shortcomings of physical life. I noticed how under the weight of a stone all kinds of repulsive worms and vermin accumulate. So it is under the weight of fear and envy when hate and thoughts of destruction come forth. Do not think that you only have to awaken what is humane and dignified in a person to make them walk the way of the light. We told you before: go your own way and let those who have chosen another path go theirs. Matter and spirit are irreconcilable opposites, as are restricting decrees and freedom. Accept from other stations only what seems acceptable to you. Those who do not want to walk with you should let you go your own way. You do not expect them to accept you, except by their own free will. Those poor people are misguided. They let themselves be fooled by beings who want to harm you. They too, will understand and in time find their way back to their true spiritual home. You, the participants in Project Sothis, should not fear anything. We are with you. Ishkumar.

[44] INIT members in Luxembourg received a message from The Seven Ethereal beings on their telephone answering machine a few days before the INIT meeting at Tarrytown, New York, in the late summer of 1996. Unlike other information from The Seven, this message arrived in English, since that would be the principal language of the meeting. They said, in part:

In the course of bygone decades, of thousands of earthly years, beings interested in the human species meet to decide on the continuation of The Project. You must not imagine that only the seven implicated in the actual development of INIT are there. No, it is a coming together of all entities interested in mankind. The interests are various. We, the Seven…, have decided to help and support the way chosen by you, in INIT. It is the way of morals, which means to understand, to acknowledge, to devise, and to act. It is not to be mixed up with religion, which means to believe. The two can be complimentary, but they are independent one from the other.

You already know that also pharisees, ghouls, swindlers, thieves, yes, even murderers, have their interested supporters here among the dead. And, as the word "higher being"—notice that we never gave us this name ourselves—does not stand, as it is often misinterpreted by falsely religious people to be purified, rid of all sin, whatever the word "sin" means for them. There are also entities here interested in that situation.

This is the seventh time that we accompany and guide you on your progress toward a free, wealthy and sane future in which humanity would have stripped off the chains of intolerance and cruelty—a future in which it will be able to establish fruitful and durable relationships with the Light, ethereal realms of existence. Our and your opponents tried to prevent this by all means.... Whatever has happened, do not lose courage. We are there. You are in the right way. You are a small number, but much depends on you and your decisions those days. We trust in you.

That ethereal voice message can be heard in its entirety on my website:

http://spiritfaces.com/3g-Msg2Wrld.htm.

Printed in the United States
138496LV00003B/1/P

9 781606 937495